AF069863

Youth Considers the Heavens

Youth Considers the Heavens

HIGH SCHOOL STUDENTS' OPINIONS ABOUT
MAN'S PLACE IN THE WORLD IN RELATION TO
THEIR ASTRONOMICAL INFORMATION

ELSA MARIE MEDER, Ph.D.
Research Associate, Bureau of Educational Research in Science
Teachers College, Columbia University

KING'S CROWN PRESS
MORNINGSIDE HEIGHTS, NEW YORK
1942

Copyright 1942 by
ELSA MARIE MEDER

PRINTED IN THE UNITED STATES OF AMERICA

Z22-DANCEY-500

King's Crown Press is a division of Columbia University Press organized for the purpose of making certain scholarly material available at minimum cost. Toward that end, the publishers have adopted every reasonable economy except such as would interfere with a legible format. The work is presented substantially as submitted by the author, without the usual editorial attention of Columbia University Press.

ACKNOWLEDGMENTS

It is with pleasure that I gratefully acknowledge the unwearied assistance of my sponsor, Professor Samuel Ralph Powers, who encouraged me and guided me in my doctoral study and who made available to me the facilities of the Bureau of Educational Research in Science. I wish to express my thanks also to Professors Irving Lorge and Herbert J. Arnold, who never failed to respond to my requests for advice and help.

My gratitude is due as well to many other professors at Teachers College for their interest, timely suggestions, and criticism given with kindness: John L. Childs, William B. Featherstone, Frederick L. Fitzpatrick, George W. Hartmann, George A. Kopp, Ralph B. Spence, and Percival M. Symonds.

My thanks are extended no less to the teachers and students who cooperated in one or another phase of this study and to the administrators of the schools in which they work, to those who helped in the clerical and statistical parts of the research, to my colleagues in the Bureau of Educational Research in Science, and to many friends and fellow-students in the Advanced School of Education.

The friendliness of those with whom I have worked and the faith and loyalty of my family have been talismans against discouragement; these have been assets for which I am deeply grateful.

E. M. M.

TABLE OF CONTENTS

1. THE PROBLEM IN ITS SETTING

    Introduction 1
    The Literature.. 2
    The Present Study 4

2. PRELIMINARY EXPLORATIONS

    A Study of Published Credos 5
    Responses to an Examination Question 8
    The Initial Attempt to Elicit Opinions 10
    Preparing a Description of Elicit Opinions from
        High School Students 12
    Responses of High School Students........................ 15

3. PREPARATION OF MATERIALS

    Preparation of Opinion "Tests"........................... 22
    *Preparing a Second Description to Elicit Opinions from
        High School Students* 22
    Estimating the Similarity of the Stimulus Situations 26
    Preparing Check Lists of Opinion 29
    Self-Rating Scale 32
    The Test Forms 33
    Preparation of Information Tests 33
    Statement of Informational Objectives 33
    Vocabulary Test 34
    Fact Test 36
    Preparation of Reading Material 36
    Premises 36

Table of Contents vii

    Generalizations 38
    "The Spangled Heavens" 39

4. COLLECTION, TREATMENT, AND INTERPRETATION OF DATA

    General Plan of the Study 41
    Administration of Tests 42
    Nature of the Scores 43
    Effect of the Order of Administration
      on Opinion Check-List Scores 44
    Definition of Most Frequently Used Statistics 44
    Differences between Pre- and Post-Test Opinion Scores 44
    Interrelations of Scores on Opinion Tests 45
    Relations of Opinion and Information Scores 47
    Relations of Scores on the Opinion Tests to Intelligence 48
    Relation between Change of Opinion and Change of Information 48

5. TREATMENT AND INTERPRETATION OF DATA PERTAINING
    TO TEST CONSTRUCTION

    Opinion Tests 51
    Validity of Information Tests 51
    Reliability of Information Tests......................... 52

6. IMPLICATIONS OF THE FINDINGS; SUMMARY

    Implications for the High School Curriculum 55
    Implications for the Education of Teachers 56
    Summary of the Study 56
      Problem and Plan of Investigation 56
      Treatment and Interpretation of Data 57
      Inferences 58

BIBLIOGRAPHY 59

LIST OF TABLES

1. Means of Ages, IQs, and Test Scores of Three Groups of Students.. 43

2. Significance of Differences between Scores on Two Opinion Check Lists for Three Groups of Students 45

3. Interrelations among Scores of Three Groups of Students on Pre- and Post-Test Check Lists of Opinion and on Pre- and Post-Test Self-Rating Scales

    a. *Correlations among Opinion Test Scores* 46

    b. *Differences among Groups* 46

4. Interrelations among Scores of Three Groups of Students on Pre- and Post-Tests of Opinion and on Pre- and Post-Tests of Information

    a. *Correlations of Scores on Opinion and Information Tests*... 47

    b. *Differences among Groups* 47

    c. *Comparison of Correlations between Scores on Opinion and Information Tests* 47

5. Interrelations among Scores of Three Groups of Students on Pre- and Post-Tests of Opinion and on Intelligence Tests

    a. *Correlations of Scores on Opinion Tests with IQ* 49

    b. *Differences among Groups* 49

6. Differences between Means of Scores of Three Groups of Students on Information Pre- and Post-Tests 49

7. Correlation of Difference in Scores on Opinion Pre- and Post-Test Check Lists with Difference in Scores on Information Pre- and Post-Tests for Three Groups of Students 50

8. Interrelations among Scores of Three Groups of Students on Pre- and Post-Tests of Information and on Parts Thereof

    a. *Correlations of Information Pretest Scores with Scores on Anchor Pretests* 53

    b. *Correlations on Information Post-Test Scores with Scores on Anchor Post-Tests* 53

    c. *Comparison of Correlations between Information and Anchor Pretest Scores and Correlations between Corresponding Post-Test Scores* 53

    d. *Correlations of Scores on Information Pretests with Post-Test Scores* 54

    e. *Comparison of Correlations between Scores on Pre- and Post-Tests of Information with Correlations between Scores on Anchor Pre- and Post-Tests* 54

    f. *Comparison of Correlations between Anchor Pre- and Post-Test Scores* 54

I

THE PROBLEM IN ITS SETTING

INTRODUCTION

Throughout his history, man has faced his problems in terms of assumptions concerning the nature of the universe and of his place in it. In all ages he has sought to explain the physical universe in which he lives. His explanations have been scientific in varying degrees and they have been constantly in a state of flux. Many of them have entered into cultural traditions and have been embodied in religious dogmas. In this manner they have become interrelated with codes of action and have become a part of a complex of beliefs and sanctions.

The moral and ethical sanctions of the Western world were developed in terms of anthropocentric conceptions of the universe which are no longer in accord with observations and thus no longer seem tenable. Because of the complex relationships existing between an individual's conception of the universe and the principles which direct his actions, emotional conflicts may be expected to arise when a person finds himself intellectually compelled to revise or discard his fundamental assumptions about the nature of the world of which he is a part.

In general, scientific research has acted to destroy many traditional ideas upon which people have based their actions. Acceptance of the findings of science may result in disintegration of the framework within which a person views his problems, leaving him without a basis for behavior. Among those who disseminate scientific findings are teachers of secondary school science. Teachers are sometimes said to be divided into two categories: one group concerned primarily with the imparting of information and the other group concerned primarily with "general education," that is, with "the fullest possible realization of [socially desirable] personal potentialities."[1] There is reason for each group to give thought to the effects of the impact of scientific information upon the ideas of the young people whom they teach. Those concerned primarily with the promulgation of information may more efficiently attain their objective if the effects of the information on the individual can be identified. Teachers concerned with the maximum development of desirable personal potentialities are committed to helping the individual build a

---

1. Progressive Education Association, Commission on Secondary School Curriculum. *Science in General Education*, p. 23. D. Appleton-Century Co., New York, 1938.

guiding philosophy, free from major conflicts, directive of attitude and action. This implies concern with the changes that occur in young people's ways of looking at life as they are brought in contact with the observations of modern science.

THE LITERATURE

The importance of the general area indicated in the introductory paragraphs is attested by writers in various fields. One of these is Lawrence K. Frank, who states:

> The need of youth and of adults is for coherent, interrelated ideas, conceptions, and meanings drawn from and solidly supported by scientific research, but presented so that they are meaningful and congruent. Without a firm conviction of the nature of the universe and man's place therein, human personality can find only a precarious foothold for the life career.[2]

Charles M. Campbell goes so far as to say that the study of beliefs is the most important as well as the most difficult task in the field of public health, for beliefs give value to life, determine in large measure the happiness of individuals and of groups, and actually have much to do with the prolongation of human life.[3] He states that the personality enters vigorously into beliefs or attitudes toward the universe of which we are a part. While as a psychiatrist he does not advocate an effort "to rob souls in trouble of their comforting beliefs,"[4] he does recommend the diffusion of scientific information, in order that people's beliefs may be in accord with what is known and may be such as to help them deal with their problems in a mature and satisfying manner rather than by an escape from reality.

This recommendation may be taken as a challenge to science teachers. S. Ralph Powers points out:

> Adolescents discover for themselves that some of their childish beliefs are inadequate . . . Science teachers have a special responsibility to help young people in such matters. They must deal with the nature of things, of the world, of human beings, of truth itself. They must be even more alert than other teachers to sense when new ideas are crowding too fast on immature minds, when there is too great incongruity between a child's past beliefs and the new ones he is meeting.[5]

---

2. Frank, Lawrence K. "The Task of General Education." *Frontiers of Democracy (The Social Frontier)*, Vol. 3, No. 24, p. 173, March, 1937.
3. Campbell, Charles M. *Delusion and Belief*. Harvard University Press, 1926.
4. *Ibid.*, p. 67.
5. Powers, S. Ralph. "The Effects of Instruction in Science on Thought, Feeling, and Action." *Teachers College Record*, Vol. 41, No. 5, pp. 411, 412, February, 1940.

Percival M. Symonds, too, found implications for science teachers in the desire of adolescents for "some set of values to which to tie themselves . . . an organization of their thinking which will give them a worthy and dignified place in the universe."[6]

In a recent study of opinions of science teachers, R. Will Burnett found that 86 per cent of the more than two thousand teachers responding to a questionnaire agreed with the statement: "Science has disrupted many of the older beliefs that man has held and cherished, therefore teachers of science have a responsibility to society to face such issues squarely in their teaching."[7] Perhaps this statement had different meanings for different individuals, for 48 per cent of the secondary school teachers also agreed with the item: "There is no conflict between science and traditional Christian beliefs." Twenty-eight per cent of the secondary school teachers indicated that they used special tact in handling the issue, "Conflicts in view of universe and man's place therein between traditional religions and science," and 29 per cent that they avoided it completely. Burnett states that "many teachers of science have felt that emotional difficulties were often the result of presenting science in such a way that it shows conflicts between faiths that the children may have."[8]

In this connection a quotation from the Yale studies of boys making the transition from secondary school to college is pertinent.

> The feeling that professors went out of their way to ridicule religion was indicated by students in several colleges. Whether this was exaggerated, a misinterpretation, or an accurate picture, the fact that the students felt this to be a deliberate attempt to discount religion is significant. It would seem that the attempt to stimulate thought and questioning too often is left incomplete. The destructive procedure seems from students' reports to be done with thoroughness without furnishing the raw materials or plan for reconstruction.
>
> Somewhat related to the attitude of the professors, and perhaps the occasions for the expressions of attitudes which upset students, are certain courses. Specifically mentioned by students are evolution courses and science courses.[9]

6. Symonds, Percival M. "Life Problems and Interests of Adolescents." *School Review*, Vol. 44, p. 506, 1936.
7. Burnett, R. Will. *The Opinions of Science Teachers on Some Socially Significant Issues: A Survey of Teacher Opinion and Its Implications for Teacher Education.* New York, 1940.
8. *Ibid.,* p. 48.
9. Hartshorne, Hugh, Ed. *From School to College,* pp. 35, 36, Yale University Press, 1939.

In one of the studies,[10] 337 sophomores were asked about the changes which had taken place in their thought. With regard to "philosophy of life," 57 per cent reported "much change" and 39 per cent reported "some change." It is noteworthy, however, that only 6 per cent felt that they had developed any new, satisfying philosophy.

The Yale studies were concerned only with boys in eastern colleges, and their implications were stated with reference to guidance in the junior college years. However, they lend emphasis to the challenge to science teachers that they accept their responsibility for helping young people achieve an intelligent and workable philosophy of living.

THE PRESENT STUDY

An individual's philosophy of living emerges from his conception of the universe and his own place in it, a conception which is in effect a highly generalized complex of intricately interrelated informations, insights, interests, attitudes, and appreciations. Changes in such a complex are exceedingly difficult to identify. The attempt to accomplish this identification is a task confronting educators in all areas. It is a special responsibility of workers in science education, for science investigation and instruction deal intimately with the nature of things.

The study reported in this volume was undertaken in an effort to identify changes of this sort in young people. It rests upon the assumption that there is some relation between ideas of the nature of the physical universe and ideas of the place of mankind in it. The teacher of science may present a description of the universe; he may even ascribe some particular place in it to man. But in the long run, greater importance attaches to the information which the students accept and to its effects on them than to the information which the teacher presents and to the manner of its presentation. Therefore it seemed desirable to try to determine whether change of information results in change of expressed opinion. Tests were prepared to measure information about the extent of the universe and other instruments were constructed to elicit statements of opinion about man's place in the world. These were given to eleventh- and twelfth-grade students before and after an experimental period during which some students continued with their usual work in physics and others carried out activities especially designed to increase their comprehension of the extent of the universe. The problem was to determine whether change of position on a test of opinion about man's place in the world was related to change of position on a test of information about the extent of the universe.

10. *Ibid.*, pp. 252, 273, 274.

II

PRELIMINARY EXPLORATIONS

An attempt to study the relationship between knowledge about the physical universe and thought about man's place in it demands as a tool a method for eliciting expressions of opinion concerning man's relation to the universe. The first step in the present investigation, therefore, was to seek such a method; this quest became a preliminary, even though subjective, study of the problem.

A STUDY OF PUBLISHED CREDOS

People do express in words their beliefs about man's place in the universe. *Living Philosophies*[1] and *I Believe*[2] are volumes of essays by noted men and women, dealing with their "convictions and beliefs concerning the nature of the world and of man."[3] It seemed plausible that an analysis of these statements might indicate something of the areas of thought which had been affected by scientific discoveries concerning the nature of the physical universe, and might reveal whether or not the authors felt that such discoveries had contributed toward the development of their points of view. Furthermore, it appeared possible that the information thus gained might be helpful in evolving a method whereby statements of beliefs might be elicited from other persons.

It was found that in many of the articles ideas of the extent of the universe and of the age and origin of the earth were expressed or implied. The portions dealing with these aspects of cosmogony were underscored, or transferred to index cards, or both. The essays were then studied with reference to these passages; during the course of this deliberation certain points became clear. These are stated and illustrated in the paragraphs which follow. Although the quotations are here necessarily given out of context, each article was considered as a whole in actually drawing the generalizations.

First, it is evident that the contributors to these books recognized and gave thought to the problem of understanding the nature of the world and of man. Theodore Dreiser writes: ". . . to be quite frank, I have thought of but little else . . . In reverence or rage or irony, as the moment or situation might dictate, I have pondered and even demanded of cosmic energy to know *Why*."[4] H. G. Wells states:

---

1. Einstein, Albert and others. *Living Philosophies*. Simon & Schuster, New York, 1931.
2. Fadiman, Clifton, Ed. *I Believe*. Simon & Schuster, New York, 1939.
3. *Ibid.*, p. ix.
4. *Living Philosophies*, p. 55.

> It has exercised my mind a lot to find out how much I could tell you of my credo in a few thousand words. Because I suppose that means telling you what I think I am, why I exist, what I think I am for, what I think of life, what I think of the world about me, and things like that. These are questions to which I have given innumerable hours — in conversation, in reading and writing, in lonely places, and particularly in that loneliest place of all, the dark stillness of the night.[5]

Hilaire Belloc refers to the attempt "to interpret ourselves and the universe about us" as the "supreme inquiry" and calls the questions raised in it "the only questions the answers to which really matter."[6]

Second, it was not easy for many of the writers to express their opinions about these questions. Pearl Buck begins her contribution with the statement: "It is no simple matter to pause in the midst of one's maturity, when life is in full function, to examine what are the principles which control that functioning."[7] Jules Romains writes: "To ask someone to tell you what he thinks of the principal problems facing man and, so far as possible, of the general nature of things, is to throw him into considerable, even painful, difficulty."[8] Thomas Mann commences his essay with the statement: "I find it singularly difficult to formulate, either briefly or in a more extended pronouncement, my philosophical ideas or convictions—shall I say my views, or, even better, my feelings?—about life and the world."[9] Irwin Edman offers reasons for such reluctance.[10] Modern philosophy, he points out, has made people conscious of the subjectivity with which they view the world, and modern psychology has made them suspect the motives underlying their beliefs. "There is indeed," he adds, "an additional reason why a contemporary hesitates, even upon invitation, to set down his credo about the universe... Every man now knows too little and too much to set down an easy ultimate."

Third, there is evidence that the knowledge these men and women possessed about the extent of the universe affected their thoughts about man's place in it. Sir James Jeans starts his essay with this paragraph:

> Quite frankly, my point of view is that of a scientist—an astronomer. In brief, this means two things. First, because I am a scientist, I am apt to see human life as a chain of causes and effects; the life of tomorrow will be what we make it today; as we sow, so shall we reap. Second,

---

5. *Ibid.*, p. 79.
6. *Ibid.*, p. 287.
7. *I Believe*, p. 33.
8. *Ibid.*, p. 213.
9. *Ibid.*, p. 189.
10. *Living Philosophies*, pp. 277, 278.

because I am an astronomer, I am apt to see the problems of today set against a background in which the whole of human history shrinks to the twinkling of an eye, and to think of these problems specially in relation to man's past history on earth.[11]

Lin Yutang states:

> Since we cannot but conceive of God as being at least commensurate with His universe, we naturally become, or I did become, awe-struck and spellbound as modern astronomy steadily revealed a wider and wider physical universe to us. The greatest enemy to old religions and to all anthropocentric faiths is the two hundred-inch diameter telescope. When I took up a New York paper a few weeks ago and read that some astronomer had discovered a new star cluster 250,000 light-years away from the earth, my notion of man's place in nature became downright ridiculous. These things are not unimportant in their bearings on our belief; they are highly important. I long ago reached the point where I realized how small and puny and humble I looked in God's, or the universe's, eyes, until the idea of a complicated system of downfall, punishment, and redemption seemed as absurd and preposterous to me as if I were to imagine myself evolving a system of punishment and redemption for a being less than the size of an ant's feeler, or even of a fair-sized maggot. We are individually not worth God's anger. We are not worth a damn, literally.[12]

Finally, it must be stated that there is wide variation among the beliefs expressed by the authors. Of course, nothing else is to be expected, for, as Bertrand Russell points out, a person's outlook on the world is "the product partly of circumstance and partly of temperament."[13] Yet two additional similarities may be noted. In commenting upon *Living Philosophies*, Sir Arthur Keith writes:

> Of the twenty-two men and women who contributed originally to these pages, only two regard the Creator—God, the One, omnipotent and personal—as having the form and properties set forth in the book of Genesis.[14]

And in discussing the contributors to *I Believe*, Clifton Fadiman states:

> These twenty-one contributors are, quite candidly, all intellectuals.

11. *Ibid.*, p. 107.
12. *I Believe*, p. 163.
13. *Living Philosophies*, p. 9.
14. *I Believe*, p. 382.

That does not mean that they are not men and women of action. It simply means that they believe in the intellect. Put them in a room together and they would disagree violently on many important matters. But on one thing they would agree; that man, at his best, is a reasoning animal.[15]

RESPONSES TO AN EXAMINATION QUESTION

The great majority of persons are less articulate than the forty-three whose "living philosophies" were studied. It seemed wise, therefore, to study the statements of other individuals before attempting to develop an instrument for eliciting expressions of opinion concerning man's relation to his world. The material selected for this consisted of a part of the final examination papers of graduate students in a course on the teaching of natural science. The requirement was: "Discuss the extent to which the concepts of science have altered traditional beliefs, world-views, and moral and ethical standards in the last two hundred years." Among the one hundred thirty papers were twenty-two which contained statements indicating something of the students' reactions to scientific concepts of the nature of the universe. The wide variation among these reactions is evident in the quotations given here.

> Science points to a purposelessness in those phenomena around which superstition and dogma grew up.

> In scientific thought and its fearless search for truth men can think for themselves and give over to God the respect that is due Him with unquestioned certainty for they will have arrived at their conclusions in logical manner.

> Judging by the recent rush back to the churches . . . one might be led to venture a guess that fewer people's beliefs and world-outlooks have been affected than we like to think.

> The Deity changes from a progenitor and becomes a skilled and talented mechanic who has built a well-ordered universe full of things beneficial to man.

> The developing scientific outlook during the past two centuries has completed the change of our cosmic picture from the geocentric universe of Ptolemy to the vast expanding universe of Einstein. It has extended the time of events from the four thousand years of Mosaic cosmogony in Genesis to the billions and trillions of years for the ages

---

15. *Ibid.*, p. xi.

of the planets and stars. It has removed God from the scene as an immediate participant in world and human events, and relegated Him to a vague expression of human aspirations, at the most.

It is true that one of the functions of science is a better understanding of the universe, but this in a restricted, material way.

Eleven of the students taking the examination made direct statements of their own beliefs. Most of these were variations of the statement: I believe that there is no conflict between science and religion. One student added:

> I do not consider that man is but a part or continuation of nature. My belief is that at a certain point in the evolutionary process, God endowed the product with an immortal soul, thus setting him apart from the rest of nature. In denying this fact I think that all moral and ethical standards, except perhaps convention, would necessarily be denied.

Several emphasized the inapplicability of scientific procedures to certain occurrences, as did the man who wrote:

> Simply because the scientific method has solved many of the mysteries of the past, mysteries to which were often attached mystic or supernatural explanations, if any, it does not warrant giving science carte blanche to attempt to explain all the mysteries which still surround us. When science attempts to delve into the sphere of the metaphysical it no longer is science, in my view. I am quite content with my own metaphysical explanations, explanations of the First Cause, for instance.

Only one person went as far as the young woman who stated:

> As a scientist this writer sees no reason for tacking on even a mathematical God with no apparent *raison d'être* outside of some soothing effects on certain internal emotional conflicts.

The impressions gained from study of these examination papers may be compared with those resulting from study of the "living philosophies." In the first place, the graduate students, like the men and women who permitted their credos to be published, seemed to be concerned with the problem of understanding man's relation to the universe. The phrasing of the examination question was not such as to require any save an impersonal treatment, and the fact that about one sixth of the students did not maintain objectivity suggests that this problem was of personal significance to them. Furthermore, it suggests that the students were on the whole less reluctant to express their opinions about the nature of things than the authors of *Living Philosophies* and *I Believe,* perhaps because they were not writing for publication, perhaps because they were more naïve. However, in the case of the students' papers there is little to warrant a conclusion that information about the nature of the universe affects thought about man's relation to it. On the other hand, there is no justification for the opposite conclusion, that such information does not affect ideas about man's place in the world.

THE INITIAL ATTEMPT TO ELICIT OPINIONS

From the analysis of some of the examination papers, it appeared that statements of opinion concerning man's relation to his world might be obtained by more direct means. In an effort to explore this possibility, the fifty-seven members of a graduate course in the teaching of science were given a sheet bearing the following statements and request.

> "Every man is occasionally visited by the suspicion that the planet on which he is riding is not really going anywhere; that the force which controls its measured eccentricities hasn't got anything special in mind. If he broods upon this . . . theme long enough, he gets the . . . idea that . . . ."

> This unfinished quotation is from an article by James Thurber.[16] The rest of the article is interesting, but this opening sentence is especially challenging. Will you allow your mind to play with the theme suggested and record the ideas which come to you?

Twenty-two students responded, eleven of whom signed their papers. Two of the respondents indicated that they considered thought along the lines mentioned in the quotation entirely futile and without either interest or importance; three others claimed that they had never speculated in such a manner. One of the three wrote:

> The idea never occurred to me, but I have heard this idea expressed by others rather frequently. This thought seems to me to bring out in these individuals:
> (1) The idea that this life is complete in itself.
> (2) A "nothing makes any difference" philosophy. Whatever I care to do and gives me pleasure is all right to do.
> (3) An attempt to rationalize and make correct for themselves to do anything which gives them pleasure.

A sixth person mentioned only thoughts concerning the physical universe,

---

16. *I Believe,* p. 295. In the papers given the students, there was no indication of the omissions.

apparently without considering man's relation to it. For example, his response included this ending to Thurber's sentence:

> ... perhaps the heavenly bodies will collide with the earth ... the sun will move close to the earth and destroy it by fire ... the oceans may swell up and cover all land surface.

All the other respondents appeared to be thinking in terms of man's relation to his world. One young woman wrote:

> We are aimless, without purpose. Perhaps our times, our civilization, our culture, are just as purposeless. We exist for a short span, and our existence means nothing in the eternity of time; the material things, the ideas, the philosophies for which we are willing to sacrifice our only connection with this earth—life, itself—in the final analysis are without value, meaningless.

Another person asked:

> Were all those other stars made just for man on this tiny earth to look at? What are they doing there? How did they get there? Do some of those suns have tiny earths like our spinning about them and is there life on them? There seems to be no beginning of time or space. Are we just hanging loose somewhere? We are here so short a time, does it pay to deny ourselves pleasures of today in hopes of others tomorrow?

In quite another vein one student wrote as his reactions:

> Maybe man is not the end of all creation. "When I consider the heavens, what art thou, O man, that the Lord should be mindful of thee?" Man should feel humble.

Another stated:

> My study of science and all that it involves has helped me to realize more clearly that there is a guiding force back of it all and that this universe is moving toward some definite goal.

Two other responses seem of especial interest, for these students referred to religion more explicitly than did the others. One man wrote somewhat impersonally:

> ... life is just a short pause on this earth; the future is indefinite. If a person is religious and believes in a life hereafter as taught by Christ, he becomes dubious of that. It brings up a real conflict in life, a search for truth, but he also realizes that he is sinning against the Holy Ghost.

Another student raised several questions pertaining to religion. He stated:

> Brooding on this theme leads one to ask, "Is there a God?" If there is a God, is there such a thing as Divine Providence? If not, religion is nothing more than an imaginative fairy tale. The horror of such a possibility often frightens one into believing this if he is not aware that a suspicion is not conclusive proof. If there is a God and He is protecting us, is Christ divine? If not, Christianity is not religion.

In general, it may be said that this exploration was valuable in two respects. First, it led to the conclusion that this type of instrument would be useful in eliciting opinions concerning man's relation to the world, that is, that even in this area, individuals would respond by stating their reactions to a given stimulus. The fact that more than one third of the papers distributed were returned was itself an indication of this, for the group was believed to include persons who resented requests to participate in experimental research, and no special effort had been made to secure returns. Second, it suggested a criterion for the preparation of instruments of this sort, for the impetus given speculation by the statement used evidently was such that thought proceeded along lines of questioning rather than of affirmation. A statement utilized to elicit opinions about man's relation to the universe would probably be effective to the extent that it provided equal opportunity for skepticism and avowal.

Because preliminary work with science teachers appeared to confirm the usefulness of the type of instrument in which a situation is described and reactions to it requested, it was decided to attempt to construct such an instrument which would be useful with secondary school students, and which would conform to the standard formulated as a result of the study with graduate students, namely, that equal opportunity be afforded for responses of different sorts.

PREPARING A DESCRIPTION TO ELICIT OPINIONS FROM HIGH SCHOOL STUDENTS

An imaginary incident involving high school boys was therefore described, and questions were based on it. The phrasing of the questions was deliberately such as to require responses in the first person, for it seemed probable that such responses would include statements which would represent the students' own opinions about man's place in nature. The description and the questions are given here.

> At a boys' camp this summer, the leader of one tent found that his boys indulged in numerous after-taps bull-sessions. He learned that one

of the boys, George, had very definite ideas concerning the nature of the universe, and was very capable in defending his views by quoting from the works of well-known scientists. Henry had equally definite ideas, which were in many cases opposed to George's. These he defended equally ably, reasoning well and quoting from the Bible and from many religious leaders to prove his views on the nature of the universe. Most of the other boys joined in the arguments, but Bob Bogart seemed to have nothing to contribute on topics of this sort, although he took an active part when the bull-sessions concerned other subjects. Apparently he did not have knowledge, ideas, or interests of any sort about the nature of the universe.

Do you think that the beliefs George and Henry hold will make any difference in the lives they lead? What difference? Why?

Is it important that Bob take an interest in this sort of problem, or is his lack of interest unlikely to affect his life? Why do you think as you do about this?

Should there be opportunity for discussion of questions of this sort so that students may learn the kind of beliefs other people hold? If you think so, what form could this opportunity take? In any case, what are your reasons for your answer?

This instrument was submitted to several experienced teachers for criticism. Three criteria for items of this sort were agreed upon: the writer's position must not be evident; there must be no weighting in favor of science; extreme positions should be presented. The group decided that the instrument under consideration met these standards, that the general form was satisfactory and that the incident described was one within the experience of high school students. However, it was felt that both the statement and the questions were too abstract. As a consequence, the incident was rewritten in dramatic form, and the questions were made more specific. This form follows.

A Midnight Conversation

On an overnight hike this summer, four boys spread their blankets on the ground, and lay looking at the stars and talking. A part of their conversation is given here.

George: Gosh, you can see a lot of stars tonight!
Henry: There must be millions of them.
Jim: And every one of them is as big as the sun.
Henry: What makes you think that?

George (speaking at the same time as Henry): No, they're bigger than the sun—most of them, I mean. The sun's just a medium-sized star, and not very bright, either, compared with some of those up there.

Jim: I wonder whether any of them have planets? Just think—millions of stars, and if each of them had eight planets like the sun . . .

George: Nine.

Jim: Nine, then, you old scientist. Anyway, there might be millions of worlds like ours. Gee, we don't seem very important, do we?

George: We're not. Mighty unimportant, if you ask me.

Henry: I don't agree with you there.

George: Well, look: how many people are there on earth, anyway?

Henry: About a billion, I guess.

George: And how many animals?

Henry: Oh, billions, especially if you count all the microscopic ones.

George: So you're just one living thing among so many other living things you can't even count them. How important does that make you?

Henry: I don't appear so, I admit. But human life is worth an awful lot, just the same.

George: Oh, I agree with you that it's fun to be alive, and I don't want to die just yet. But heck, it doesn't mean anything! Us here—and up there, hundreds of thousands of light-years away, millions of suns. What good are we anyway? Why, I heard my father say that if something should happen to the sun so that the whole solar system disappeared, it wouldn't make any difference to the universe.

Henry: But it would. Look, you've left God out of all this. He made the earth, and made us, too, and even if we can't see what His purpose is, He has one.

George: I don't see how you can think that. There are so many stars, and the universe is so enormous, that I just don't see any point in our being here.

Henry: Sure, the universe is enormous, and everybody feels that way sometimes. You know David said, "When I consider thy heavens, O Lord, the work of thy fingers, the moon and the stars, which thou hast ordained, what is man that thou art mindful of him, and the son of man that thou visitest him?"

Jim: That's right. Say, you know the Bible, don't you?

Henry: There's something else in it that sort of fits in here. That's the place where it says that . . .

Bob (speaking for the first time) : Aw, why don't you shut up? Or else talk about something worth talking about!

From this conversation, it is apparent that George and Henry each have some definite ideas about the nature of the universe, and that their ideas are very different. With their discussion in mind, answer the following questions:

What difference, if any, will the ideas George and Henry have make in their actions? Consider such things as the way they treat their parents, their school work, their attitude toward people of other races and of other economic levels, church-going, dating girls, getting a job, and so forth.

Why do you suppose that Bob had so little to say? If it is true that his silence was due to lack of information on the subject, what sort of information do you think he should be given?

Do you think that there should be opportunity for boys and girls to discuss questions of this type so that they may learn the kind of beliefs others hold? What are your reasons for your answer? What form could such an opportunity take?

RESPONSES OF HIGH SCHOOL STUDENTS

This instrument was distributed among high school students of the ninth, tenth, eleventh, and twelfth grades, and about one hundred thirty responses to each question were studied, especially with respect to the kinds of opinions represented. These responses came from schools in several states.

To the first question, that concerning the effects of the boys' ideas upon their lives, a large number of responses included value judgments. In the majority of instances these judgments favored Henry rather than George. The following are examples of this tendency.

> Henry would be a person who did a lot of sensible thinking.
>
> Henry is by far the more serious and the more sensible of the two.
>
> George doesn't seem to have a wholesome trust in any matters.
>
> I think Henry is certainly the more logical of the two.
>
> Henry seems to know the proper attitude.

A small number of the students indicated approval of George.

> To me George seems to be a very intelligent boy. He doesn't speak without some thought. He seems to be very studious and enthusi-

astic. He seems to me to be a type of a boy to serve as an example for all young boys of his age.

Henry believes what is told him as a fact, but George is suspicious of things told by other people and therefore analyzes things for himself from a realistic standpoint. George will try to better himself by using his initiative and ambition to look into things that don't sound exactly right, while Henry will usually be in a rut by following other people.

In answer to the first question, also, a number of the young people stated or implied more directly their own opinions about man's relation to the world. Some quotations illustrate this.

I feel the way Henry does; God has a reason for us here—though we may seem unimportant. If we do feel unimportant as George, we won't care about our studies or our friends or religion. These are very serious things not to care about. The most serious of these is our religion and when we lose that then we really have the wrong attitude. If I didn't have a church then I would feel unimportant like George.

I think Henry is very right about God putting us on this earth for a reason.

It seems as though George is looking toward the evolutionists' side. Henry appears to me to be more of a Christian so far as faith in our Lord is concerned . . . It is true to a certain extent what George said about life, "But heck, it doesn't mean anything!" but to each and every individual life is priceless and I know that a person that enjoys living and tries his best does not want to pass away.

The universe is probably much more important but still I agree with Henry that human life is important.

[George] has forgotten about his creator, God, who put him on this earth.

God made the earth and people for a purpose and He also made the universe.

George, I believe, is probably the better informed . . . For when God was mentioned as making the earth he scoffed at the idea, being no older than the others, because he could not conceive of God or any other man creating all the other planets. Probably he as much as the rest had been taught that God did create the earth but at the same time he had his own ideas upon the subject.

In answering the second question, the boys and girls generally gave as reasons for Bob's silence his possible ignorance, lack of interest, low intelligence, laziness, or weariness. Some considered him immature, as did the boy who wrote:

> He probably is not as grown up as his friends, as he doesn't realize the value of the discussion.

Some felt that Bob may have been embarrassed.

> Bob may have been a more sensitive boy. He may be embarrassed to air his opinions of the more profound subjects and he may feel embarrassment for his friends. Many boys are afraid to make fools of themselves like this. I do not think that this feeling should be encouraged.

> Bob may have felt that it was too large and too close a question to discuss, that nothing could be gotten from such a discussion.

Others thought that he recognized that nothing could be decided by the discussion or that he wished to keep the other boys from getting into a serious argument.

> He didn't want to get mixed up in a conversation which would get him nowhere.

> Maybe Bob had the same opinion as Henry and to keep the conversation from becoming an argument he didn't say anything.

A few thought that Bob really was in agreement with George, and some thought that he was afraid to face facts.

> I think that Bob had so little to say because he sort of agrees with George in some things.

> I think that Bob had so little to say because he knew George was correct in his theories (or ideas).

> I think that Bob was smart in saying nothing because he knew how unimportant he was but tried to live in a way as to think he was the only one really important.

> I think Bob had plenty of information but he was lacking the courage to look at things and think about himself and what his purpose is. There are some people, you know, who shun to think about and analyze what they are here for. They are actually afraid to think.

I don't think Bob ever thought about those things, and when he did I think he was scared and afraid to look at things that way.

By far the majority of the students recommended that Bob study both science and religion. Two illustrations will probably suffice.

If Bob's silence on this subject was due either to indifference or lack of knowledge, he should be given both religious and scientific information. He should be given information pertaining to the hypotheses of creation of both the world itself and man. He should be informed of the regularity by which this universe works. He should be made to realize the immense size of it and the great number of living things on it. After he has gained all the information he can on the subject, he should combine the ideas he has gotten from his religious and scientific information and base his convictions on these.

He should be given more information on the sciences and religion. He should have information on sciences because practically everything around him deals in science. He should have more knowledge of religion because without this he would more or less be in the same predicament that George is.

As these papers were written in science classes, it is scarcely surprising that a number of students recommended in rather general terms the study of science. A few, however, were more specific. One boy wrote:

I would recommend what George thinks, namely, that we are such an insignificant part of the universe, so why did God put us here along with the billions of other animals?

Another, who did not sign his paper, and who had included in his answer to the first question the statement, "If you think of the world being made scientifically, there is no God," wrote tersely and emphatically:

I would give Bob information on how the universe was made *scientifically*.

A number of students emphasized the importance of religious instruction, as did those quoted here.

He should have some instruction on religion as an answer for our purpose on earth.

He needs religion and an ambition or aim for which to prepare.

Bob should be given information regarding the Bible and he should

> be taught to believe in God so that he could have more faith in himself and in mankind than George.
>
> He should be told that we are important, that we must be or we wouldn't be here, because God made everything and everybody for a purpose.
>
> Bob should be shown that the ultimate source was a God that had to begin the work of the universe. He should be shown that the Bible affects him and other people and his relations with them. He should be taught that a study of science produces deeper and better thought.
>
> I do not think it necessary to further his information on the subject as I believe he would probably live just as good a life never thinking about it as he would if he were to delve very deeply into the matter. This is disregarding the religious end of the subject, as I believe that everyone should have a thorough knowledge of the Bible and religion.
>
> I don't think it necessary to teach Bob about astronomy because I don't think it is important, but I do think that he should have some view on why people live. This cannot be taught him but he must make up his own mind on the subject from hearing it discussed.
>
> It would be well for him to study various of the better philosophers, Confucius, Plato, Aristotle, perhaps Bacon and Emerson, and *certainly* the Bible.

A few students wandered from the question. Some of their remarks may be of interest. One girl wrote:

> I think that Henry was right to recite some scriptural verses because I would have referred to that also. Maybe not the same verse but something similar. I also would have told him that we should never question God. He made the universe and he knows why he made it that way so we just have to be satisfied. God put us here for a purpose so we surely have something to live for.

A boy stated:

> In this world of turmoil there is no room for people who do not look to God for some consolation. In life there is only a quantity of days to look forward to, but a sacrilegious person will hate himself when death grows near.

There was greater agreement among the students with regard to the third question, that concerning the desirability of young people holding discussions of this sort, than with respect to either of the other two. Only eight of the one hundred thirty respondents answered in the negative; four of these gave as reasons that the discussion would get nowhere. One boy wrote:

> A discussion of this type might be fun, but it would be futile.

Another reaction is illustrated by the statement:

> I do not believe that I would care to discuss these kinds of questions because they are not particularly interesting.

One girl wrote in explanation of her negative answer:

> I feel that people who believe in God will do so unless and until they are disillusioned by something done by their church or synagogue. People who do not believe in religion will continue not to. I have noticed that such discussions only lead to discomfort and friendship breaking. The only thing that can be done is to give adequate science courses, and then, if people can still believe in God, more credit to them!

The explanations given by the students who answered the question affirmatively fall into several categories. One sort of reason was to the effect that by finding out the beliefs of others tolerance would be attained; another that in this way one's own thinking would become clearer. For example, one boy wrote:

> Discussions of this type would help us to be more tolerant of others and perhaps prevent more conflict of the type now going on. They would also prevent warped concepts of the world about us and thus provide a balanced outlook on life.

A different kind of reason was expressed by the boy who stated:

> I do believe boys and girls should have an opportunity to discuss questions like this but not to find out the beliefs of others. I think this would be a very bad idea. A person might lose many friends just because he does not believe as they do. I think the good part of one of these discussions is to enable a person to present his facts and then fight for them.

A small group of students saw in such discussions an opportunity to change other people's ideas. Two examples follow.

> If the person had the wrong viewpoint you could correct him and therefore give him a better slant on life.

It might encourage those who lack the qualities of Henry to form an aim in life, have a religion, and live to help and to be helped by others.

Many students gave more specific reasons than these. "My reason is," said one anonymous respondent, "that I *would like very much* to have such a discussion." (The italics are the student's.) Some signed answers were:

> Well, this may sound silly but I think many are puzzled with the question.
>
> I think many are troubled by the problem of just how important their own life is and whether it should be used just for fun, or is it worth while to try to make it useful.
>
> Most of the students have a sort of wanting to know about vast fields of which their knowledge is limited.
>
> The question has always been puzzling.
>
> There are a hundred and one questions that I have no answer for.

In answering this question, some students pointed out that such discussions were quite natural.

> It is not a matter of "should there be"; there is such discussion going on every day between students and friends.
>
> There should be opportunity for such discussions, and obviously there is, as there is nothing unusual about some boys talking while sitting round a campfire on a summer hike.

In one school, the feeling of the students concerning the desirability of discussing questions such as that raised by the test item was found to persist for several months. They had responded to "A Midnight Conversation" early in the month of October. Late in January, the teacher asked them to indicate what area of study they felt would be most important for them to pursue further; he reported that the one response which was repeated again and again was "A Midnight Conversation."

In general, the results of the trial of the instrument appeared to warrant an attempt to use it more objectively and under more controlled conditions.

III

PREPARATION OF MATERIALS

If the method of evoking statements of opinion about man's place in the universe developed in the preliminary explorations was to be used as a tool in a controlled study of students' opinions, more than one stimulus was needed; a second description, comparable to "A Midnight Conversation" was essential. Furthermore, the opinions stated in response to either stimulus must be susceptible to numerical description. In other words, two forms of an opinion "test" must be prepared. Two forms of an information test were obviously also required in a study of a problem stated as being to determine whether change of position on a test of opinion about man's place in the world is related to change of position on a test of information about the extent of the universe. Nor does this problem statement only imply the necessity of having measures of opinion and of information; it also implies the importance of assuring an opportunity for increase of information in that area. Three kinds of material were therefore prepared: tests of opinion, tests of information, and a pamphlet containing astronomical subject matter.

PREPARATION OF OPINION "TESTS"

Preparing a second description to elicit opinions from high school students: In order that students might be given two different stimuli to which to react, an attempt was made to describe a situation which would elicit responses similar to those called forth by "A Midnight Conversation." As a first step in doing this, several brief statements were composed, which were presumed to epitomize the responses of the students to the questions asked about this conversation. A few statements were added which were more extreme than any actually made by the students. The list prepared is given here.

Boys and girls do well to discuss the deeper questions of life.

Talking about subjects such as these helps one to think about them more clearly.

Topics such as "The Nature of the Universe" and "The Meaning of Life" are so vast that to talk about them makes one's thinking confused.

Young people should let such deep questions alone.

Boys and girls in high school should accept without question what older people whom they respect tell them about philosophical questions.

I have never thought about things like "Why are we on earth?"

Sometimes I think about the value of life and similar topics.

My own position is somewhere in between Henry's and George's, but it is nearer to George's than to Henry's.

Sometimes I feel the way George does, but I don't let my mind dwell upon it very long.

I am in thoroughgoing agreement with George—I feel that there is no point in our being here.

I occasionally feel that ideas like George's are true, but I hope that they are not.

I think Bob was the most sensible one of the group.

I am in complete agreement with Henry.

My own position is somewhere in between Henry's and George's but it is nearer to Henry's than to George's.

Henry's position is correct as far as it goes; however, I would state it more strongly.

George should study religion so that his ideas will then become more like Henry's.

George and Henry should forget such matters and spend their energy on something more practical.

Bob should study both science and religion.

Bob should learn to accept the unimportance of the earth and its inhabitants in the vast universe.

Bob should be taught that religion is the only truth, and that the "facts" of science do not help a person to live a good life.

Henry should be helped to outgrow his immature beliefs.

These boys should be let alone, for what they think about these questions will make no difference in the kind of lives they lead.

These statements were used as guides in describing a second situation, that is, the description was written for the purpose of eliciting responses similar to these. The second incident is recorded here.

After the Rehearsal

The chorus and orchestra of Duncan High School were preparing for a joint concert. As it was to be given on a Sunday, they decided to end the program with a hymn. The conversation reported here took place among one group of students as they left a rehearsal.

Larry began to whistle the hymn they had just sung, and the others joined in, the boys whistling and the girls singing lustily.

> The spacious firmament on high
> With all the blue ethereal sky,
> And spangled heavens, a shining frame,
> Their great Original proclaim:
> The unwearied sun from day to day
> Does his Creator's power display;
> And publishes to every land
> The work of an almighty hand.
>
> Soon as the evening shades prevail,
> The moon takes up the wondrous tale;
> And nightly to the listening earth
> Repeats the story of her birth;
> Whilst all the stars that round her burn
> And all the planets in their turn
> Confirm the tidings as they roll,
> And spread the truth from pole to pole.
>
> What though in solemn silence, all
> Move round the dark terrestrial ball,—
> What though no real voice nor sound
> Amid their radiant orbs be found,—
> In reason's ear they all rejoice,
> And utter forth a glorious voice,
> Forever singing as they shine,—
> "The hand that made us is divine."

Dave: Hey, we'd better not make so much noise. After all, it's late, and lots of people are asleep.

Jean: Oh, they shouldn't mind our singing a hymn. Now, if we were singing "Roll Out the Barrel" or something like that!

Dave: I guess it wouldn't make much difference to someone who was waked up.

Jean: Well, I feel more like the hymn than like "Roll Out the Barrel" right now, anyway. Golly, what a night!

Lois: It's a pretty good hymn at that. It's Haydn's music, you know.
Larry: And the words are pretty good, too. That's just the way people feel on a night like this.
Lois: The heavens certainly are spangled, aren't they?
Mac: The heavens are spangled, all right, but I don't see where that "does the Creator's power display."
Dave: There's no doubt that the science in the hymn could be improved. That last verse surely makes the earth the center of things, doesn't it?
"What though in solemn silence, all
Move round the dark terrestrial ball!"
Jean: That's quite a phrase—"dark terrestrial ball." I'm not even sure what it means.
Lois: Well, there's one value of taking Latin, anyway. *Terra* means earth, so it's the dark earth ball.
Dave: But, of course, the earth isn't dark. It shines just like the other planets.
Lois: Well, part of it's dark. But, as you said, it isn't the center of things. The hymn has the sun and the stars and the planets all going around the earth.
Mac: Of course, I guess we have to give the man who wrote the poem poetic license and let him mess up science if he wants to, but I think he's got the wrong idea back of it.
Dave: What do you mean, Mac, "He's got the wrong idea?"
Mac: Oh, all those lines, like "Their great Original proclaim" and "The hand that made us is divine." That's foolish. Those stars up there aren't telling *me* that the hand that made them is divine!
Larry: Sure they are! When you think of it, it must be God who made them.
Dave: Wait a minute, Larry. What *are* the stars telling you, Mac?
Mac: Well, for one thing, how big the universe is. When you think how big they are and how terribly far away, you get some idea of how enormous the universe is. And then, they tell me that the earth isn't important. It can't be, because it's just a little planet going around a star that's smaller than lots of those stars.
Larry: But the earth is important. If it isn't, what are we doing here?
Mac: I'm sure I don't know. Do you?
Larry: No, but God put us here, so we must be here for some purpose.
Mac: That's what you *want* to think. You like to think that, and you like to think God made the stars and the planets and that they tell us about "their great Original." I'd rather be honest. I know that people

aren't important. Gee! if everybody died, or if the whole earth disappeared, or all the planets, or even the sun, what difference would it really make? All the rest of the stars would still be there—stars bigger than the sun!

Larry: I think you're wrong—about our unimportance, I mean.

Mac: Look at the stars. Choose one, any one. Now imagine the sky without that one. Does it matter if it's not there?

Larry: No, but that just goes to show that we're important, because it would matter if the sun went out and the earth came to an end.

Mac: Why?

Larry: Why? Because it's all in God's plan. He made us and He has a reason for us here. If He didn't, we wouldn't be here right now. And when He wants it to, the sun will disappear and the end of the world will come. But we can't tell what His plans are. Why should we? It's enough to be here!

Mac: That doesn't make sense to me.

Lois: I still think it's a pretty good hymn. Let's sing it again.

Estimating the similarity of the stimulus situations: Some estimate other than the subjective judgment of the writer was desired concerning the similarity of "After the Rehearsal" to "A Midnight Conversation" as stimulus situations for evoking opinions concerning man's relation to the universe. As a first step in securing this, a list of statements of possible reactions to each situation was prepared. In the case of "A Midnight Conversation," the list was based upon the series of statements recorded on pages 22 and 23. The earlier statements were scrutinized, some were revised according to certain of the criteria of Thurstone and Chave, some were retained, and others were discarded.[1] The standards selected were:

> As far as possible, the opinions should reflect the present attitude of the subject rather than his attitudes in the past.
>
> The material should be edited so that each opinion expresses as far as possible only one thought or idea.
>
> Each opinion should preferably be such that it is not possible for subjects from both ends of the scale to endorse it.
>
> As far as possible the statements should be free from related or confusing concepts.

[1] Thurstone, L. L. and Chave, E. J. *The Measurement of Attitude*, pp. 56-58, University of Chicago Press, 1929.

Preparation of Materials

In the case of "After the Rehearsal" the statements were paraphrases of this new list. The parallelism in the two series is evident.

"A Midnight Conversation"

I do not agree completely with either George or Henry.

My ideas are more like George's than like Henry's.

I am in complete agreement with Henry.

I am in thoroughgoing agreement with George.

I am more inclined to agree with Henry than with George.

Henry's position is correct as far as it goes; however, I would state it more strongly.

I am undecided as to whether George or Henry has the right ideas.

It doesn't matter to me whether or not there is any point in our living.

I believe that everyone on earth is here for some purpose.

I feel that there is no meaning in human life.

I believe Henry to be too positive in his statements.

Henry should be helped to outgrow the beliefs he expressed.

Henry should learn to accept the unimportance of the earth and its inhabitants in the vast universe.

These boys should be let alone, for what they think about these questions will make no difference in the kinds of lives they lead.

George and Henry should forget such matters and spend their energy on something more practical.

George should study religion.

George should be taught in such a way that he will come to agree with Henry.

George should learn that religion is the only truth.

George should be taught that the "facts" of science do not help a person to live a good life.

"After the Rehearsal"

In my opinion, neither Larry nor Mac was entirely right.

My own position is nearer to Mac's than to Larry's.

I agree thoroughly with Larry.

I am in complete agreement with Mac.

My own opinion is more like Larry's than like Mac's.

My ideas are somewhat similar to Larry's, but I would be even more positive than he is.

My ideas are sometimes like Larry's and sometimes like Mac's.

I do not care whether or not there is a purpose to life.

I am sure that there is real meaning in human life.

I believe that human life is without purpose.

Larry states his opinions too strongly, for he cannot prove them.

Larry should recognize that human beings and the earth they live on are of little or no importance in the vast universe.

The opinions Mac and Larry expressed are actually unimportant, for they will make no difference in the kinds of lives the boys will live.

Mac and Larry should stop thinking about questions that have no answers and think instead about more practical matters.

Mac especially should think very seriously about religious matters.

Mac's education should be directed in such a way that he will come to see that Larry's ideas are correct.

Mac should be taught that there is only one source of truth and that that is religion.

Mac should learn that living a good life is unrelated to knowing scientific "facts."

As a second step in obtaining an estimate of the similarity of the stimuli, the statements were typed on cards, and each series together with the corresponding descriptions was given to judges. The judges were requested to arrange the cards so that the statements were in descending order according to the extent to which they appeared to express opinions of a religious nature. The rankings of each judge were correlated with those of every other judge, according to the Spearman rank-difference formula. The correlations between judges were considered as between comparable tests, and their average was substituted in the Spearman-Brown prophecy formula.

In the case of "A Midnight Conversation" the rank-order correlations for three judges were .90, .92, and .88; applying the Spearman-Brown formula, *r* equalled .96. This indicates that a combination of three equally good repetitions of judgment would correlate to the extent of .96 with a combination of three other equally good repetitions of judgment. The rank-order correlations for the judges who ranked the items for "After the Rehearsal" were .94, .95, and .89; in this case *r* equalled .97 after substituting in the Spearman-Brown formula.

As the third step in estimating the similarity of the descriptions, the ranks given by the judges to each item were added together, and new ranks were assigned on the basis of these sums. These new ranks for the corresponding items on each of the two series of statements were then compared; the rank-order correlation was .97. This seemed to indicate, although in a somewhat roundabout fashion, that the two conversations could be considered as comparable.

Preparing check lists of opinions: Given two comparable situations which could be used to evoke the expression of opinions concerning the nature of the world and of man, the problem next was to prepare two comparable check lists of opinions to which students could be asked to respond. The statements which had been used for estimating the comparability of the situations were not considered suitable for this purpose, because these were couched in terms of comparison of the respondents' opinions with those of the boys in the descriptions.

Sixty-two eleventh-grade students were given copies of "A Midnight Conversation," with this conclusion:

> From this conversation, it is apparent that George and Henry both have some definite ideas about the nature of the universe, and that their ideas are very different. It also appears that they have both given some thought to questions which may be termed "deep" or philosophical, that is, questions such as, "What is the meaning of life?" Please state as briefly as you can what you consider their ideas to be.

Space was left for the response, and then there was the request:

> Please state your own ideas. Do you agree with George or Henry, or with neither one? How would you express your opinion?

From the response to this request, sixty-four statements of opinion were obtained. These statements were duplicated on slips of paper and complete sets were distributed to judges. The judges were directed to arrange the slips so that the statement on top of the pile was the one which expressed most clearly the opinion that man was dominated by supernatural forces, and the statement on the bottom of the pile was that which expressed most strongly the opinion that man was not so dominated.

The first ten judgments returned were recorded. The ranks assigned to each item by each of the ten judges were added together, and new ranks were given in accordance with these sums. Each judgment was then correlated with the combined judgment according to the Spearman rank-difference formula. The correlations ranged from .57 to .90.

The two sets of judgments for which the coefficients were smallest were discarded. The ranks assigned to each item by each of the remaining judges were added together and a second reranking was made on the basis of these sums. The eight sets of judgments were correlated with the new ranks, and r was found to range from .61 to .88. When the Spearman-Brown prophecy formula was applied, the value of r was found to be .97. This was taken as an estimate of the reliability of the scale.

To obtain an estimate of the reliability of each item, the squares of the differences between the ranks assigned by each judge and the rank assigned on the basis of the combined judgments were added together. The fourteen items for which the sum was largest, that is, the fourteen least reliable items, were discarded, and the remaining fifty statements were assigned consecutive ranks. These were converted into linear scores according to the technique described by Hull,[2] a method which assumes the ranks to be along a continuum and to represent a normal curve.

Each item could now be described in terms of two numerical values, a scale score which varied from .7 to 9.3 and a reliability estimate which varied from 136 to 1,662. These values were used as coordinates, and a chart was constructed indicating the position of each item. Two series of twenty-four items each were prepared by selecting items at fairly even intervals along the axis of the chart representing the opinion scale, and insofar as possible from among the lower values along the axis representing their reliability. The items having scale scores of 8.2 or above and of 2.4 or below were included in both series; in all, forty statements were used. The sum of the opinion scale values for one set of statements was 116.6; for the other set, this was 118.2. The sums of the reliability estimates were 15,492 and 15,518, respectively. The two series were therefore considered as comparable opinion scales.

The two sets of statements, together with the opinion scale score of each item, are given here. A high score indicates that the statement expresses the

---

2. Hull, Clark L. *Aptitude Testing,* pp. 382-390. World Book Co., Yonkers, N. Y., 1928.

idea of man as dominated by supernatural forces; a low value indicates that the statement expresses the opposite idea.

Series A

9.3 God had a reason for making us and all the earth.
8.6 God has a purpose for putting us on earth.
8.2 God made us to live until He thinks we are unfit to live.
7.8 God made us and the universe.
7.4 All human life was created by God.
6.9 God will reveal His plan for our lives.
6.6 We should go through life without questioning God.
6.4 There must be some supernatural power controlling all the planets.
6.0 Everyone is here to carry out part of a plan.
5.7 We must take God's work into consideration.
5.4 We all mean something under God.
5.1 The bigger the world, the bigger its purpose must be.
4.6 Human life is not valuable to the universe.
4.4 The sun, not the earth, is the most important part of the universe.
4.0 The world can get along without us.
3.8 We have a lot to learn about the universe and about life.
3.7 The earth exists for man.
3.3 That we are here proves we are important.
2.6 There is no reason to think that the world was created by God.
2.4 The stars, the planets, and the sun are important to us.
2.1 Human life is more important than all the rest of the universe.
1.8 While we make up a very small part of the universe, we are still very important.
1.4 Human life is valuable and important.
0.7 We are the most important of all living things.

Series B

9.3 God had a reason for making us and all the earth.
8.6 God has a purpose for putting us on earth.
8.2 God made us to live until He thinks we are unfit to live.
7.8 God is the head of our universe.
7.2 God decides who is to be on this earth and who is not to be on this earth.
7.0 God created heaven and earth.
6.7 God has a purpose for everything.

6.5 It is not always possible to know God's will.
6.1 We are put on this earth for a trial before we enter heaven.
5.6 Human beings were put on earth for a purpose.
5.3 Religious concepts are more likely to be true than scientific ideas.
5.0 The "facts" about the universe have nothing to do with the meaning of human life.
4.3 It is wrong to consider ourselves unimportant.
4.1 Human beings are unable to judge whether living is worth while or not.
3.9 I just cannot see any reason for being here.
3.4 It does make a difference whether or not we are on earth.
3.3 That we are here proves we are important.
3.1 Human life does not *appear* to be worth very much, but it is.
2.8 There is a lot worth living for.
2.4 The stars, the planets, and the sun are important to us.
2.1 Human life is more important than all the rest of the universe.
1.8 While we make up a very small part of the universe, we are still very important.
1.4 Human life is valuable and important.
0.7 We are the most important of all living things.

Self-rating scale: The reactions to "A Midnight Conversation" expressed by the respondents in the preliminary study were often stated in terms of agreement or disagreement with one of the boys described. It seemed desirable, therefore, to afford an opportunity for those who responded to the refined instrument to express their opinions by comparison with the opinions of the characters in the stimulus incident. To secure such an expression in a form which could easily be scored, it was decided to ask each respondent to mark his own position along a line on which the positions of two of the characters described had been indicated. The request took the following form.

It is possible to think of people's ideas on a subject as lying along a line. From the conversation, we see that George's ideas and Henry's are rather far apart; on a line, that might be shown in this way:

By reading the next diagram, we find that person A agrees with George; B agrees with Henry; C feels even more strongly about the matter than Henry; and persons D and E have ideas somewhere in between those of Henry and those of George, although D's ideas are somewhat like George's.

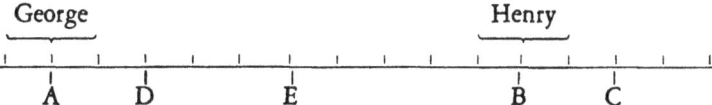

Use the next line to show where your opinion should be placed.

The test forms: Each form of the completed opinion "test" consisted of a description of a conversation purported to have taken place among young people, a request for a brief identification of the two chief characters described, a check list of opinions, and a self-rating scale. The descriptions are given on pages 13-15 and 23-26. The request for identification of the characters in "A Midnight Conversation" is recorded on page 29; when the incident was "After the Rehearsal," the names Mac and Larry were substituted for George and Henry in this and on the self-rating scale. The check list consisted of one of the series of statements arranged in random order; the directions which preceded it on the test follow.

> Read the statements below and place a check in the parentheses after each statement with which you agree, thus (∨). Then reread the statements, and place a cross in the parentheses after each one with which you disagree, thus (X). If you cannot decide whether you agree or disagree, put a question mark in the parentheses. Be sure that you mark every statement in one of these ways.

As two descriptions and two check lists had been prepared, four test forms were available by using each series of statements with each incident.

PREPARATION OF INFORMATION TESTS

Statement of informational objectives: To test information concerning the extent of the universe, it was considered desirable to have two measures, one specifically of knowledge of vocabulary, and the other directed toward more general knowledge. As a first step in the preparation of these tests, informational objectives were stated. The broad generalizations upon which these objectives were based follow:

> The earth is the astronomical body from which man views other celes-

tial objects; it is a part of the solar system. The sun is one of billions of stars in our galaxy; there are uncounted galaxies.

Seven objectives were stated, namely:

An understanding of the earth as an astronomical body, from which man views the other celestial objects

Knowledge that the solar system includes eight planets other than the earth and thousands of planet-like objects

Some understanding of the nature of comets and meteors

An understanding of the sun as the star around which the earth and other objects in the solar system are moving

An understanding of the stars as members of a galaxy

Some information concerning the nature of galaxies

Knowledge of scientific theories of the earth's origin

The second step in preparing the tests was to outline the information pertaining to these objectives. Within any one of them, several different kinds of knowledge may be recognized. For the present purpose, five sorts were identified—knowledge of vocabulary; other factual knowledge, both in isolation and in its relationships; knowledge of methods by which facts become known; and knowledge of people's varying beliefs and ways of thinking. The information pertinent to each objective was classified as one of these sorts of knowledge. The classification was carried out in such a way as to result in a chart of thirty-five rectangles. Each objective was represented by a row, and each type of knowledge by a column. Notations were made in each rectangle indicating appropriate information. This procedure not only facilitated the classification of information but simplified the task of sampling the material for the tests.

Vocabulary test: The words listed in the column of the chart headed "Vocabulary" were chosen after a study of indexes and glossaries of high school and college textbooks of general physical science and of popular and elementary astronomy books. Multiple-choice test items were written for each of the one hundred thirty words selected. The usefulness of these items was judged in two ways. The first was the subjective judgment of three persons, and sixty items were discarded upon this basis. The second judgment was based upon estimates of the difficulty and of the validity of each item.

Test items which are most discriminatory with respect to the members of a group are those to which half the persons tested respond correctly; these items have a difficulty level of 50 per cent. One criterion of acceptability of items for use in this study was a difficulty level between 5 and 95 per cent.

The criterion of the validity of an item was the extent to which it measured what the entire test measured. The correlation of the score on the entire test with the response, correct or incorrect, to an item—the Pearson biserial correlation—was taken as an index of the validity of the item. Valid items are those having high positive biserial correlation coefficients. For the purposes of this study, the lower limit of a "high" coefficient was placed at that correlation coefficient corresponding to the value of Fisher's z-function which is 2.5 times its standard error. In this instance, 2.5 times the standard error of z was .1689. By definition, then, no item was valid which had a biserial correlation coefficient smaller than .17.

The biserial correlation coefficients of the seventy test items under consideration were estimated from the responses of 222 high school seniors by a technique described by Flanagan.[3] The coefficients obtained ranged from −.10 to +.65; the difficulty levels ranged from 12 to 98 per cent. Fifty-eight of the seventy items were considered valid, having biserial correlation coefficients of .17 or greater; sixty-eight of the items were within the accepted range of difficulty.

The vocabulary test was envisaged as requiring fifteen to twenty minutes of student time; this indicated that each of the two forms to be prepared should include thirty to forty items. The necessary number of items was reduced by the plan of using certain items on both forms—"anchor" items. The primary reason for employing this technique was a desire to investigate the question: Does the relation between scores on an anchor test and total test scores differ before and after instruction?

Fifty-four items, whose difficulty levels ranged from 16 to 93 per cent passing, were chosen by discarding all items having biserial correlation coefficients lower than .17 and the four having coefficients closest to this value.

Eighteen items were used in both forms of the test, and half of those remaining were used in each form. The thirty-six items in each of the test forms were arranged in order of difficulty, the easy ones being placed first. As an estimate of the comparative validity of the forms, the biserial correlation coefficients for the items were added together. The sums were 16.39 and 16.29; the averages were .455 and .453. To compare the difficulty of the test, the proportions of successes were added together. The sums were 1,854 and 2,113; the average per cents were 52 and 59.

3. Flanagan, John C. "General Considerations in the Selection of Test Items and a Short Method of Estimating the Product-Moment Coefficient from Data at the Tails of the Distribution." *The Journal of Educational Psychology*, Vol. 30, pp. 674-680, 1939.

Fact test: The test to measure knowledge of a more general character than vocabulary was prepared in much the same manner as the vocabulary test. The information which had been indicated on the chart was sampled, and multiple-choice items were written to test this information. After the items had been judged on a subjective basis, a preliminary test of one hundred twenty-five items was administered to ninety high school seniors, and the analysis was carried out as in the case of the vocabulary test. The same standards were postulated. In this instance, 2.5 times the standard error of z was .2680; the correlation of which this z is a function is .2618. Items with biserial correlation coefficients of .26 or greater were therefore declared to be valid. There were four negative coefficients; the positive coefficients ranged from .05 to .85, and of these one hundred eight were sufficiently high for the items to be considered valid. All the items were within the accepted range of difficulty.

As about thirty minutes were allotted for responding to the fact test, sixty items were used in each form. Again, certain items were included in both forms. In this case, the items used twice were selected in such a manner as to constitute two comparable cycles of fifteen items each. It was planned to see whether the relation between the scores on the matched items differed before and after instruction, as well as to investigate whether there was a change in the relation of the scores on the anchor items to those on the entire test. To facilitate item arrangement, a chart was prepared, using as coordinates the numerical values by which each item could be identified. Items with very high correlation coefficients were placed in the anchor cycles. The similarity of the two cycles with respect to validity is evident from the sums of the biserial r's of the items: 10.52 and 10.51, averaging .701 in each case. The sum of the item difficulty scores was 7,756 in both cases, or an average difficulty level of 52 per cent. The two nonanchor cycles, each containing thirty items, were also comparable. The sums of the biserial r's of these items were 12.18 and 12.59; the averages were .406 and .420. The sums of the difficulty estimates were 1,260 and 1,420; both averaged 42 per cent.

Each form of the test, when assembled, contained the two anchor cycles and one of the nonanchor cycles. The order of the items was randomized.

PREPARATION OF READING MATERIAL

The first steps in the preparation of reading material for the subjects of the study were clarification of the premises for selecting the generalizations to be developed, and the choice and statement of these generalizations.

Premises: Facts are important, for they provide materials with which the mind may operate; words, too, are important, for they facilitate thought and expression. But if facts remain unorganized and unused, if words have sound but shallow meaning, the time spent in learning them might better have been otherwise employed. If facts are interrelated and if words express these interrelationships, then knowledge may lead to insight, and insight to generalized attitudes. But knowledge of words and facts and interrelationships is in itself no insurance against confusion and unhappiness; other kinds of knowledge offer more in this regard. It is important that an individual know how the facts became a part of human culture, know how facts were learned. This involves some understanding of the method of science, as a powerful tool, as a rigorous standard of thought, and as a self-corrective procedure. Because the scientific method involves action upon inadequate bases and revision of judgment when new evidence is available, the conclusions of science are not to be accepted as dogma. Hypotheses wholly in accord with observation may later be shown to be untenable. Change is inherent in the method of science; flexibility is the factor which has kept science living. Scientific discoveries have never been universally welcomed, and it was inevitable that the progress of science should be impeded by those who feared its effects upon the beliefs they cherished.

Three kinds of knowledge, then, are goals of learning—knowledge of facts, knowledge of method, and knowledge of the history of thought, for these can contribute to those skills and attitudes involved in the development of a scientific world-picture. Study directed toward achieving increased comprehension of the extent of the universe involves all three of these. Some of the major scientific findings in this area may be briefly summarized.

The universe consists of millions of galaxies, most of which are now thought to be receding from one another. Our sun is situated in one of these galaxies, an average star among many thousand million stars. The solar system consists of this star, its nine known planets and their satellites, at least a thousand planetoids, numerous comets, countless meteors, and rarefied matter. Evidence indicates that it came into being several thousand million years ago. Millions of years after its origin, conditions on earth were favorable to the development of life; after further eons of time, man emerged from simple beginnings. His curiosity prompted him to investigate the world in which he found himself; he set up explanations of his observations and transmitted them to his progeny. In the course of a long cultural development, he has formulated and discarded explanations and beliefs; he has discovered and, to some extent, applied the "scientific method" of unbiased inductive and deductive reasoning from carefully collected data; he has attempted to differentiate between natural laws and emotional longings; he is learning to recognize the variability and diversity and interrelatedness of all phenomena.

Generalizations: Proceeding upon this basis, a list of generalizations was prepared. This list was discussed with a number of experienced science teachers and was reviewed in connection with a survey of astronomy textbooks, textbooks for science survey courses, and popular and semipopular writings on astronomy. From a revision and extension of the list, certain generalizations were selected to be developed in the reading materials.

> The earth is the astronomical body from which man views the other celestial objects; it is a part of the solar system.
>
> The solar system includes nine known planets, their satellites, planetoids, comets, and meteors; all these bodies revolve around the sun.
>
> The sun is a gaseous body whose diameter is about 110 times that of the earth. It constantly radiates enormous amounts of energy, of which the earth receives a small fraction.
>
> The sun is one of billions of stars in our galaxy, the only one from which our distance can be conveniently expressed in miles. For other stellar distances, the unit used is the light-year.
>
> The stars are not unchanging, but seem to undergo a life cycle, which is evident to us by the differences in brightness of stars whose distances and densities are known. The sun appears to be somewhere between maximum brightness and extinction.
>
> The galaxy of which our sun is a member (not the central star, however) has been found to be a disk with a short diameter measured in thousands of light-years and a long diameter measured in tens of thousands of light-years.
>
> There are uncounted galaxies. Study of those distant more than 100 million light-years has shown them to be grouped by the hundreds into supergalaxies.
>
> Despite the vast numbers of heavenly bodies, the universe is comparatively empty. Interstellar distances are enormous; intergalactic distances are virtually incomprehensible.
>
> The kind of matter observed when studying other astronomical bodies does not appear to differ significantly from that which is known on earth.
>
> The methods of science are those of systematic, critical examination and are opposed to blind reliance on authority.

Scientific methods include the formation of generalizations from hypotheses which were set up to explain observed phenomena or were deduced from previous knowledge, and which were tested, revised, and confirmed in the light of accumulated data.

Man has always attempted to explain and control the universe in which he finds himself. Many of his theories have been in accord with the observations which could be made.

Scientists have not been content with the explanations of the universe that they inherited, but they have continued to study its nature and extent. We still know only in part.

"The Spangled Heavens": These generalizations were used as guides in writing a pamphlet of fifty-four pages, which was entitled *The Spangled Heavens.*[4] The generalizations concerning methods of thinking and the incompleteness of present knowledge were developed in connection with astronomical information. The material included in the pamphlet is indicated to some extent in the following outline:

INTRODUCTION. Why people ask questions. Questions that can be answered. Questions that no one can answer. Questions that only you can answer.

CHAPTER I. HERE WE ARE! BUT WHERE ARE WE? Getting down to earth. We charter a rocket ship. Mountain heights and ocean depths. How to measure the diameter of the earth. Spinning at 800 miles per hour. How we know that the earth rotates. How we know that the earth revolves. The earth's companion. A model of the earth, the moon, and the sun.

CHAPTER II. "OTHER WORLDS THAN THIS"— Worlds without moons. A world with two moons. Worlds with many moons. "Life on other worlds"? Combinations of atoms. Living things and temperature. Living things, oxygen, and water. Can there be life on Venus? Can there be life on Mars? Martian canals.*** Discovering "new" worlds. Planets predicted and later found. A planet predicted but never found. The case of the missing planet.*** Comets that come back and comets that disappear. Comets' heads and comets' tails. If the earth and a comet should meet. Falling meteors. The rocket ship again. The sun's temperature. The energy the earth receives. The energy the earth does not receive. Where the sun's energy comes from.

4. Powers, S. Ralph and Meder, Elsa M. *The Spangled Heavens.* Bureau of Educational Research in Science, Teachers College, Columbia University, New York, 1941.

CHAPTER III. "AND IT CAME TO PASS" – Beginning in the midst of things. When was the "beginning"? How people change their ideas. The nineteenth-century notion of the "beginning." The sun a star. Twentieth-century ideas of the "beginning." Are there other solar systems?

CHAPTER IV. "AMID A CROWD OF STARS" – The Dipper's bowl. Counting 100 billions. The temperature of stars. The size of stars. What mass is. How stars may change with time. The age of stars. The Milky Way galaxy. The place of the sun in the galaxy. The "emptiness" of the galaxy. Rapid motions.

CHAPTER V. WHEN WE COME TO THE END OF THE MILKY WAY – Millions of galaxies. The shape of galaxies. Galaxies are moving. Real speed. Supergalaxies. What do we know?

In manuscript form the pamphlet was evaluated in terms of two criteria. The first of these was that of scientific accuracy; the second that of readability. Success in attaining these criteria was estimated by discussion with competent persons who critically read the material, and in the case of the second criterion, by application of the "Lorge Formula for Estimating Grade Placement of Reading Materials."[5] The "Readability Index" found upon applying this test was equal to 5.7; since this indicates the grade level of the reading material, it appeared unlikely that it was too difficult for high school students.

5. Lorge, Irving. "Predicting Reading Difficulty of Selections for Children." *The Elementary English Review,* Vol. 16, pp. 229-233, October, 1939.

IV

COLLECTION, TREATMENT, AND INTERPRETATION OF DATA

GENERAL PLAN OF THE STUDY

The testing instruments and the teaching instrument appeared to be in a form suitable for objective study of the problem. Three groups of eleventh- and twelfth-grade physics students participated in an experiment lasting for fourteen school days. All the subjects in all the groups were given the tests on the first two and last two days of the experimental period; the ten intervening days were spent differently by each of the three groups. During this time, one group studied *The Spangled Heavens,* reading the material and answering the questions asked in it, but holding no class discussions until after the experimental period. The students in the second group engaged in such activities as their teachers preferred for the purpose of developing an enlarged comprehension of the physical universe and its vastness; there was free class discussion. The third group served as control; these students continued with their usual work in physics.

The classes which studied *The Spangled Heavens* were asked to refrain from discussion in order to secure a better measure of control over their classroom experiences. In a foreword to the pamphlet, the students were offered an explanation of the request that they study it. The foreword is reproduced here.

> *Why you are asked to study this pamphlet.* Two years ago this conversation took place between the teacher in charge of a high school study hall and a student who was regarding a science book with distaste.
>
> *Student:* This book is terrible.
> *Teacher:* Why, what's wrong with it?
> *Student:* The words the author uses! He ought to write in English.
> *Teacher:* Doesn't he?
> *Student:* Well, listen—"Certain species may be caused to reproduce by vegetative propagation in order to avoid Mendelian segregation."
> *Teacher:* Whew!
> *Student:* I'd like to tell him about it!
> *Teacher:* Why don't you? He probably would appreciate a letter.

Student: What's the use? The book is brand new, and they probably won't change it for a long time. By that time the man who wrote it might even have lost my letter.

This conversation suggested to the teacher that books should be tried out before they are printed. Consequently, this pamphlet is in multilithed form instead of in printed form. We are asking a number of students in approximately a dozen high schools throughout the United States to study it as it is now, and to give us their suggestions for its improvement. We hope to use their suggestions when we prepare the booklet for the printer.

There are a number of questions in the book which ask directly for your help, for instance, "Is there anything in this chapter that you do not think is explained well? What?" We hope that you will answer these questions thoughtfully. From our point of view it is also important that you answer the other questions—the usual sort of textbook questions—equally thoughtfully. Otherwise we cannot tell whether a question is a good one or a poor one. (Of course, from your point of view, too, it is important to answer these questions carefully.)

We have tried to write a pamphlet that is interesting and informative. We hope we have succeeded well enough for you to enjoy studying it. We are quite sure that you can help us make it better. Of course, we shall thank you in print for your cooperation, but, as there is a chance that you may never happen to see that acknowledgment, we take this opportunity of thanking you for the help you will give us.

The students who participated in the second group were taught in whatever way their teachers deemed the most effective from their knowledge of themselves and of their students. Some of these teachers requested and received copies of *The Spangled Heavens* to use as reference material. All of them were given copies of the generalizations used in writing the pamphlet as a definition of the area within which they were to teach.

ADMINISTRATION OF TESTS

An attempt was made to minimize any possible effects of the order of test administration upon test scores. The order of the pretests was therefore varied, and the tests used at the close of the experimental period were given in an order the reverse of that used at its beginning. In an effort to lessen the effects of the differences between the forms of the tests, each form was used as the pretest for some students and as the post-test for others.

NATURE OF THE SCORES

On the opinion tests two scores were obtained for each student, one on the check list and one on the self-rating scale. The check-list score was determined by averaging the scale scores for each item accepted, a procedure which gives more reliable results than that of averaging the scores of the rejected items.[1] The self-rating score was estimated by comparison with a ruler having ten equal divisions between the ends of a line as long as that on the students' sheets.

TABLE 1
Means of Ages, IQs, and Test Scores of Three Groups of Students

	"T" GROUP Mean	n	"P" GROUP Mean	n	"C" GROUP Mean	n	ALL GROUPS Mean	n
Age in months	212.0	95*	209.2	134	212.0	112*	210.9	341
IQ†	99.8	117	103.2	134	104.1	113	102.4	364
Check list								
pretest	5.03	117	4.82	134	4.88	113	4.91	364
post-test	4.95	117	4.68	134	4.92	113		
Self-rating								
pretest	5.58	111‡	4.92	127‡	5.57	102‡	5.33	340
post-test	5.67	111‡	5.13	128‡	5.42	106‡		
Vocabulary								
pretest	45.0	117	36.7	134	46.5	113	42.4	364
post-test	53.3	117	54.4	134	56.1	113		
Facts								
pretest	44.2	117	38.8	134	52.2	113	44.7	364
post-test	78.4	117	78.5	134	54.9	113		

* The report of the students' ages was incomplete.
† Otis Higher Examination, with the exception of 21 cases in the "T" group, where the New Stanford Achievement Test was used.
‡ Several students did not indicate their positions on the self-rating scale.

On the vocabulary tests three scores were obtained for each respondent, one on the anchor items, one on the nonanchor items, and the sum of these. Because there were four choices for each item, each score was computed as three times the number correctly marked minus the number incorrectly marked. The fact tests were treated similarly; as there were two sets of anchor items, there were four scores rather than three for each individual. The sum of a student's total score on the vocabulary test and of his total score on the fact test was considered as a measure of his information. The separate anchor and nonanchor scores were used in a comparison of the relation of the anchor tests to the total test before and after instruction; this is reported in Chapter V.

Table 1 affords an indication of the nature of the test scores, of the ages

1. Lorge, Irving. "The Thurstone Attitude Scales. I. Reliability and Consistency of Rejection and Acceptance." *The Journal of Social Psychology*, Vol. 10, pp. 187-198, May, 1939.

and intelligence quotients of the students, and of the number of subjects in each group. The "T" group is the group of students whose teachers planned their activities; the "P" group refers to the students who studied the pamphlet; and the "C" group is the control group.

EFFECT OF THE ORDER OF ADMINISTRATION ON OPINION CHECK-LIST SCORES

It was hypothesized that the order of administration of the tests would not affect the scores on the opinion tests. This hypothesis was tested on the one hundred seventy-six subjects whose pretests were scored first. Eighty-two of these students had taken the opinion pretest before the vocabulary and fact pretests; the remaining ninety-four students had had the information pretests before taking the opinion pretests. As a measure of the divergence from the hypothesis, the statistic t was used, that is, the ratio of the difference between the means of the samples to the best estimate of the standard error of this difference. This ratio was found to equal 0.58339, a value which, according to a table of the sampling distribution of t, is known to occur 50 to 60 per cent of the time, and is consequently insignificant. The hypothesis was therefore accepted.

DEFINITION OF MOST FREQUENTLY USED STATISTICS

In calculating Pearson product-moment correlation coefficients, the number of estimates of the parameters was taken as the number of subjects in each group, not as the number of classes to which the students belonged; that is, the subjects were considered as having been selected at random from among physics students in urban high schools. The coefficients for the three groups of students are recorded in tables throughout this chapter. The 5 per cent level of significance was accepted.

To test for the significance of the difference between two correlation coefficients, the difference between their z-functions was assumed to be normally distributed. In certain tables in this and the following chapter, the differences are expressed as the difference of the teacher-directed group from the control group, of the pamphlet group from the control group, and of the pamphlet group from the teacher-directed group. Obviously, the last of these is related to the other two. Obviously, also, this could be expressed as the differnce of the teacher-directed group from the pamphlet group by a reversal of the signs.

DIFFERENCES BETWEEN PRE- AND POST-TEST OPINION SCORES

The first question to be answered in an investigation of a problem involving determination of the relation between change of position on a test of opinion and change of position on a test of information is this: Is there a change of position of the group score on the test of opinion? To answer this question, the statistic *t* was used to test the significance of the difference between the pretest and post-test check-list scores for each group of classes. As is apparent from Table 1, the means on the two occasions differed. Table 2 shows that the difference, in the case of the students using the pamphlet, is statistically significant on the 2 per cent level, that is, it would occur in fewer than 2 per cent of the samples from a population with a parameter equal to zero. The difference in the case of the other groups is not statistically significant. The question raised, then, must be answered in the affirmative for the pamphlet group, in the negative for the teacher-directed and for the control groups. During the course of the experiment, the opinions of the students in the pamphlet group moved away from a position exemplified by a character in each test form and described for convenience by the statement: "Man is dominated by supernatural forces."

TABLE 2

Significance of Differences between Scores on Two Opinion Check Lists for Three Groups of Students

	"T" GROUP	"P" GROUP	"C" GROUP
Value of t	1.50657	2.52860	0.74207
Per cent of random samples in which t would occur	10-20	1-2	40-50

INTERRELATIONS OF SCORES ON OPINION TESTS

The question, Is there a change of position of the group score on the opinion scale? suggested the question, Is there a relation between pre- and post-test scores of opinion? The fact that the first question was answered affirmatively for one group of students and negatively for the others raises two additional questions: Do the groups differ from one another with respect to their opinions at the beginning of the experiment? Do they differ from one another in this regard at its conclusion? Answers to these may be found from a study of the correlations among the opinion scores. The coefficients of correlation calculated were those between the pre- and post-test check-list and self-rating scores. The data pertinent to the questions raised are summarized in Table 3, page 46.

A relation between the pre- and post-tests of opinion was found both in the case of the check-list scores and in the case of the self-rating scores. In both instances the coefficients of correlation are highest for the control group, lowest for the teacher-directed group. In the case of the correlation between the self-rating pre- and post-test scores, the control group differed significantly from the other two groups. In the case of the correlation between scores on the pre- and post-test check lists, the teacher-directed group differed significantly from the other two groups. It may be said, therefore, that the students most likely to place themselves in the same position on the post-test as on the pretest were in the control group, and that the students least likely to accept the same check-list statements on the post-test as on the pretest were in the teacher-directed group.

A difference between the groups at the beginning of the experiment was evidenced by the correlations between the check-list and self-rating pretest

TABLE 3

Interrelations among Scores of Three Groups of Students on Pre- and Post-Test Check Lists of Opinion (A and B, respectively) and on Pre- and Post-Test Self-Rating Scales (K and K′, respectively)

3a. CORRELATIONS AMONG OPINION TEST SCORES

	r_{AB}	$r_{KK'}$	r_{AK}	$r_{BK'}$
"T" group	.54	.54	.05†	.02†
"P" group	.70	.59	.40	.48
"C" group	.73	.81	.29	.38
All groups	.67	.66	.26	.31

† Not significant.

3b. DIFFERENCES AMONG GROUPS; VALUES OF z

	DIFFERENCE OF "T" FROM "C"	DIFFERENCE OF "P" FROM "C"	DIFFERENCE OF "P" FROM "T"
r_{AB}	−.32*	−.06	.26*
$r_{KK'}$	−.53*	−.46*	.07
r_{AK}	−.25	.12	.38*
$r_{BK'}$	−.38*	.13	.50*

* Significant.

scores. The scores were significantly related in the pamphlet and control groups but not in the teacher-directed group. A similar difference among the groups was apparent at the close of the experiment. In other words, in the teacher-directed group the students' opinions as evidenced by their acceptance of statements of opinion were not likely to be the same as the opinions they declared themselves to hold by indicating their position on a continuum. In the other groups, the two declarations were likely to be of the same general character.

RELATIONS OF OPINION AND INFORMATION SCORES

The questions hitherto raised were susceptible to answer in terms of the opinion tests alone. This was not the case with the next question: Is there a relation between scores on the opinion and on the information tests? The answer to this, as will be seen, was in the affirmative when the entire sample of students was considered, and as a result two further questions were asked, namely:

Does the relationship differ among groups?

Does the relationship differ on initial and final administration of the tests?

The correlations useful in answering these three questions were those of the opinion pretest scores with the information pretest scores and of the opinion post-test scores with the information post-test scores. The opinion test scores used were obtained by averaging the check-list and self-rating scores; the information test scores were the sums of the scores on the parts of the test. Table 4 summarizes the pertinent data.

TABLE 4

Interrelations among Scores of Three Groups of Students on Pre- and Post-Tests of Opinion (A' and B', respectively) and on Pre- and Post-Tests of Information (C and D, respectively)

4a. CORRELATIONS OF SCORES ON OPINION AND INFORMATION TESTS

	$r_{A'C}$	$r_{B'D}$
"T" group	−.15*	−.09
"P" group	−.40*	−.30*
"C" group	−.07	−.10
All groups	−.23*	−.17*

* Significant.

4b. DIFFERENCES AMONG GROUPS; VALUES OF z

	DIFFERENCE OF "T" FROM "C"	DIFFERENCE OF "P" FROM "C"	DIFFERENCE OF "P" FROM "T"
$r_{A'C}$	−.80*	−.36*	−.28*
$r_{B'D}$	.01	−.20*	−.21*

* Significant.

4c. COMPARISON OF CORRELATIONS BETWEEN SCORES ON OPINION AND INFORMATION TESTS

	$z_{A'C} - z_{B'D}$
"T" group	−.06*
"P" group	−.12*
"C" group	.04
All groups	−.06*

* Significant.

The correlations between scores on the opinion and on the information tests were significant and negative in every instance for the total sample of students in the study. It may be said, therefore, that the more information the students possessed, the lower their scores on the opinion test, both in the check-list and in the self-rating form. A low score on the opinion tests represents a position closer to an opinion that man is unaffected by supernatural forces than to one that man is dominated by such forces.

Only in the pamphlet group were both the pretest and the post-test correlations between opinion and information scores significant. This fact must be recognized in considering the results of the experiment. Students in the pamphlet group differed from students in the other groups initially with respect to the correlation between opinion and information. They were subjected to different treatment than the other students—they studied *The Spangled Heavens*. Their opinions were significantly different at the end of the experiment from their opinions at its beginning.

The third question asked at the beginning of this section remains to be answered. Except in the case of the control group, as Table 4c shows, the correlations of opinion with information scores on the pretests differed from the correlations of opinion with information scores on the post-tests.

RELATION OF SCORES ON THE OPINION TESTS TO INTELLIGENCE

The relationship between the scores of opinion and the scores of information suggested the question, Is there a relationship between scores on the opinion test and scores on an intelligence test? and suggested also its corollary, Are there differences among the groups with respect to this relationship? Table 5 indicates the answers. The correlations between the scores on the opinion tests (averages of check-list and self-rating scores) and those on the intelligence tests were significant and negative for the pamphlet group and also for the entire sample of students. The pamphlet group differed significantly from the teacher-directed and control groups with respect to these correlations.

The questions raised may be answered by stating that students with a higher I Q were more likely to have a low than a high score on the opinion tests, that is, their expressions tended to be nearer the opinion that man is not under the control of supernatural powers than to the opinion that he is under such control. The group using the pamphlet differed from the other groups with respect to this relationship.

RELATION BETWEEN CHANGE OF OPINION AND CHANGE OF INFORMATION

The next question which arose was this: Do the students who exhibit the

TABLE 5

Interrelation among Scores of Three Groups of Students on Pre- and Post-Tests of Opinion (A' and B', respectively) and on Intelligence Tests (E)

5a. CORRELATIONS OF SCORES ON OPINION TESTS WITH IQ

	$r_{A'E}$	$r_{B'E}$
"T" group	−.15*	−.08
"P" group	−.23*	−.18*
"C" group	−.07	−.06
All groups	−.16*	−.11*

* Significant.

5b. DIFFERENCES AMONG GROUPS; VALUES OF z

	DIFFERENCE OF "T" FROM "C"	DIFFERENCE OF "P" FROM "C"	DIFFERENCE OF "P" FROM "T"
$r_{A'E}$	−.08*	−.17*	−.09*
$r_{B'E}$	−.02	−.11*	−.11*

* Significant.

greatest change of opinion also evidence the greatest change in information? Two groups of students participating in this experiment were provided with opportunities to acquire information about the extent of the universe. For each group of students the means of the scores on the vocabulary and fact tests may be ascertained from Table 1, page 43; these indicate that the students in the pamphlet group initially possessed less information than those in the other groups and that they acquired more information during the experiment. Table 6 shows the differences in these means for the groups.

When these differences are considered in conjunction with the differences between the opinion scores for the three groups, recorded in Table 2, page 45, the fact which stands out is that the pamphlet group, which made the greatest gain in information, is the group which changed significantly in opinion. This suggests that there is a relation between change of position on the opinion test and change of position on the information test, a relation in which an increase in information is associated with change of opinion away from the position that man is dominated by supernatural powers and toward the position that man is not under such domination.

An attempt was made to confirm the existence of this relationship. The

TABLE 6

Differences between Means of Scores of Three Groups of Students on Information Pre- and Post-Tests

	"T" GROUP	"P" GROUP	"C" GROUP
Vocabulary	8.3	17.7	9.6
Fact test	34.2	39.7	2.7

differences between scores on the opinion pretest check list and on the post-test check list were correlated with the differences between scores on the information pretest and on the information post-test. As Table 7 shows, the correlation coefficients were not significantly different from zero for any group. Of course, it must be remembered that the difference between two scores is less reliable than either of the scores; the correlation between two such differences cannot, therefore, be very reliable.

TABLE 7

Correlation of Difference in Scores on Opinion Pre- and Post-Test Check Lists with Difference in Scores on Information Pre- and Post-Tests for Three Groups of Students

"T" group .... .10
"P" group .... .02
"C" group .... .03

Apparently the relation between change of information and change of opinion cannot be considered as proved. As indicated earlier, students in the pamphlet group differed initially from students in the other groups with respect to the correlation between opinion and information scores. The change of opinion which the pamphlet-group students evidenced may have been related to the factor which caused them to differ in this manner. But even if it were not related to this factor, it does not follow that the change of opinion was related merely to change of astronomical information. These students acquired their information from *The Spangled Heavens.* There may have been something there, perhaps the exhortation of the introduction, which caused them to change their opinions. The other students who gained a substantial amount of information were under the direct influence of their teachers, a fact which may have caused these students to maintain, rather than to change, their opinions. None of these factors can be entirely ruled out, and as a consequence it is not possible to conclude that there is a relation between change of opinion about man's place in the world and change of information about the extent of the universe, although it is possible to infer that such a relation exists.

V

TREATMENT AND INTERPRETATION OF DATA PERTAINING TO TEST CONSTRUCTION

OPINION TESTS

The chief difficulty in the preparation of the opinion tests was found to be the establishment of a satisfactory criterion of validity. If an opinion is a verbal expression of an attitude, and an attitude may be defined as a tendency to act, a satisfactory criterion of validity might be the actions of the persons who express opinions. However, it would be a well-nigh impossible task to observe an adequate sampling of the actions of students whose responses are used in constructing an opinion test. Furthermore, even overt actions may not really represent attitudes. In this study, validity was assumed on the basis of the method and care used in preparing the opinion tests. There are two arguments for the validity of the check lists: they were composed of statements made by students from samples comparable to that with which the tests were used, and there was agreement among judges deemed competent as to the scores to be assigned to the items. There seemed to be less justification for assuming the validity of the self-rating scale.

In their test of attitudes toward the church, Thurstone and Chave included a graphic scale on which the subject indicated his estimate of his attitude.[1] Three reference points were given, namely, strongly favorable to the church, neutral, and strongly against the church. These investigators found a correlation of .67 between scores on the attitude scale and positions on the self-rating scale. As may be seen from Table 3a, page 46, the correlation between the opinion check-list and self-rating scores in the present study is of the order of .3. The difference between this coefficient and that found by Thurstone and Chave may have resulted from the fact that their subjects were in college and graduate school. However, it seems more probable that the difference may be traced to the complexity of a self-rating process which requires the respondent to compare his own position with the position of others.

VALIDITY OF INFORMATION TESTS

The problem of the validity of information tests is less difficult than that of the validity of opinion tests, for biserial correlation coefficients may be

---

[1]. Thurstone, L. L. and Chave, E. J. *The Measurement of Attitude*, pp. 78, 80. The University of Chicago Press, 1929.

used as indexes of validity of the individual information test items. The information tests used in this investigation were composed of items with high biserial correlation coefficients, and "anchor" sets of items were included in each form. These anchor sets may be considered internal criteria of the validity of the test as a whole. The correlation of scores on an anchor test with scores on the total information test constitutes an index of the validity of the total test. In this instance there are four such indexes for the pretest and four for the post-test, as the scores on the total test were correlated with those on the vocabulary anchor test, on two different fact anchor tests, and on the sum of these three. Consideration of these correlations made it possible to answer the question raised during the construction of the tests: Does the relation between scores on an anchor test and total test scores differ before and after instruction? Pertinent data are recorded in Table 8, parts a, b, and c.

The coefficients are somewhat inaccurate, because the anchor scores are components of the total score. Nevertheless, the magnitude of the correlations—in no instance less than .68 for a single group—may be taken as indicating the validity of the total test in terms of the anchor tests as criteria. It is evident that the order of magnitude is the same for the post-tests as for the pretests. Table 8c shows the differences between the correlations of anchor test with total test scores before and after the experimental period for each of the three groups of students. In only one instance was there a significant difference. In general the relation between the scores on the anchor tests and on the total test was the same after instruction as it was before instruction.

In the determination of biserial correlation coefficients for use as a measure of item validity, the factor of item difficulty is involved. The difficulty of an item, whether judged subjectively by a test respondent or objectively by the test scores of a group, is different after instruction than before, absolutely if not relatively. Apparently, however, the use of biserial correlation coefficients as indexes of item validity is justified. The biserial correlation coefficients of the items used in this investigation were computed from the scores of students who had not received specific instruction in the subject matter of the test. The items were used with students before and after they had received such instruction. It was found that the anchor items measured what the test measured both before and after instruction. The value of biserial correlation coefficients as indexes of item validity was not affected by changes in item difficulty resulting from instruction.

RELIABILITY OF INFORMATION TESTS

The question was raised as to whether any of the anchor tests could be used in place of the total test. To answer this question, each pretest score was correlated with the corresponding post-test score, and the coefficients for the separate tests were compared with one another and with the coefficient for the total test. The results are reported in Table 8, parts d, e, and f, page 54. The significant differences in Table 8d are evidence that fifteen or eighteen items are too few to secure reliability. However, in no group did the pre- and post-test correlations of the forty-eight-item test differ significantly from those of the total test, which contained ninety-six items. And, as Table 8f shows, although in no case is there any significant difference between the pre- and post-test correlation coefficients for the shorter anchor tests, in most cases the coefficients for the shorter tests differ significantly from those for the forty-eight-item test. It would appear that the forty-eight-item test, made up as it was of items whose biserial correlation coefficients had an average value of approximately .6, was sufficiently long for reliability.

TABLE 8

Interrelations among Scores of Three Groups of Students on Pre- and Post-Tests of Information (C and D, respectively) and on Parts Thereof: Fact Anchor Set 1 (L and L'), Fact Anchor Set 2 (M and M'), Vocabulary Anchor (S and S'), and Combined Anchor Sets ($W = L + M + S$ and $W' = L' + M' + S'$)

8a. CORRELATIONS OF INFORMATION PRETEST SCORES WITH SCORES ON ANCHOR PRETESTS

	r_{CS}	r_{CL}	r_{CM}	r_{CW}
"T" group	.75	.72	.81	.93
"P" group	.79	.83	.79	.93
"C" group‡	.68	.73	.82	.94
All groups§	.74	.76	.79	.93

‡$n = 99$ §$n = 350$

8b. CORRELATIONS OF INFORMATION POST-TEST SCORES WITH SCORES ON ANCHOR POST-TESTS

	$r_{DS'}$	$r_{DL'}$	$r_{DM'}$	$r_{DW'}$
"T" group	.81	.83	.78	.94
"P" group	.74	.80	.82	.95
"C" group‡	.70	.80	.77	.94
All groups§	.74	.80	.78	.95

‡$n = 99$ §$n = 350$

8c. COMPARISON OF CORRELATIONS BETWEEN INFORMATION AND ANCHOR PRETEST SCORES AND CORRELATIONS BETWEEN CORRESPONDING POST-TEST SCORES; VALUES OF z

	$z_{CS} - z_{DS'}$	$z_{CL} - z_{DL'}$	$z_{CM} - z_{DM'}$	$z_{CW} - z_{DW'}$
"T" group	.14	.29*	.06	.05
"P" group	.13	.09	.06	.19
"C" group	.03	.17	.14	.02

* Significant.

TABLE 8 (Concluded)

Interrelations among Scores of Three Groups of Students on Pre- and Post-Tests of Information (C and D, respectively) and on Parts Thereof: Fact Anchor Set 1 (L and L'), Fact Anchor Set 2 (M and M'), Vocabulary Anchor (S and S'), and Combined Anchor Sets ($W = L + M + S$ and $W' = L' + M' + S'$)

8d. CORRELATIONS OF SCORES ON INFORMATION PRETESTS WITH POST-TEST SCORES

	r_{CD}	$r_{SS'}$	$r_{LL'}$	$r_{MM'}$	$r_{WW'}$
"T" group	.75	.56	.58	.55	.72
"P" group	.81	.68	.62	.58	.79
"C" group	.82	.65	.70	.68	.80

8e. COMPARISON OF CORRELATIONS BETWEEN SCORES ON PRE- AND POST-TESTS OF INFORMATION WITH CORRELATIONS BETWEEN SCORES ON ANCHOR PRE- AND POST-TESTS; VALUE OF z

	$z_{CD}-z_{SS'}$	$z_{CD}-z_{LL'}$	$z_{CD}-z_{MM'}$	$z_{CD}-z_{WW'}$
"T" group	.34	.31	.36	.08†
"P" group	.29	.39	.46	.04†
"C" group	.40	.31	.34	.08†

† Not significant.

8f. COMPARISON OF CORRELATIONS BETWEEN ANCHOR PRE- AND POST-TEST SCORES; VALUES OF z

	$z_{SS'}-z_{LL'}$	$z_{SS'}-z_{MM'}$	$z_{LL'}-z_{MM'}$	$z_{WW'}-z_{SS'}$	$z_{WW'}-z_{LL'}$	$z_{WW'}-z_{MM'}$
"T" group	.03	.02	.04	.26*	.24	.28*
"P" group	.10	.16	.07	.25*	.35*	.42*
"C" group	.09	.06	.03	.31*	.23	.26

* Significant.

VI

IMPLICATIONS OF THE FINDINGS; SUMMARY

The study reported here is appropriate in the light of the propositions that people seek to comprehend the physical universe around them; that an individual's conception of the universe influences his philosophy of living; and that it is a responsibility of education to help each person develop for himself a philosophy which "takes account of human welfare, is free of superstition, misconception, and paradox, and is in accord with scientific knowledge of human beings and the universe of which we are a part."[1] It is by no means the only study whose appropriateness could be similarly stated. In this investigation only change in astronomical information was considered. It is equally probable that people's opinions concerning man's place in nature may be affected by information in other areas; for example, by information about the great age of the earth and the development of living things on it or about the biological and psychological nature of man. Knowledge of history, acquaintance with the great religious and philosophic systems, familiarity and concern with current social problems, all may influence people's opinions about man's relation to the universe. Despite the limitations of this study, however, the findings have certain implications for education, especially with respect to curriculum construction.

IMPLICATIONS FOR THE HIGH SCHOOL CURRICULUM

Ordinarily astronomical material is not studied in the senior high school. Most junior high school science courses contain at least one unit on the earth and its relation to the solar system, but customary courses in the tenth, eleventh, and twelfth grades give no place to this or to related subject matter. If there is a relationship between an individual's conception of the universe and the principles upon which his actions are based—and it is the fundamental assumption of this study that there is—surely information bearing upon an understanding of the universe should not be omitted from the science curriculum at a time when, according to their own testimony, boys and girls are seriously considering questions of a philosophical nature. On the contrary, the curriculum should include material which will help young people to develop a world-picture in accord with modern science, one within

---

1. Powers, S. Ralph. "The Effects of Instruction in Science on Thought, Feeling, and Action." *Teachers College Record*, Vol. 41, No. 5, p. 412, February, 1940.

which they may build for themselves philosophies of living which will be serviceable and emotionally satisfying.

The pamphlet used in this study was apparently an appropriate means of supplying information which young people can utilize in their consideration of philosophical problems. The fact that those who studied the pamphlet without teacher intervention evidenced changes in opinion while those who were under the direct influence of the teachers did not alter their opinions suggests that the former group felt less constrained than did the latter. There may be occasions when impersonal presentation of information is advantageous. Other textual material, similar in nature to *The Spangled Heavens,* may be valuable in high school classes, that is, pamphlet material written simply and directly for the purpose of developing a few generalizations selected upon explicit premises.

IMPLICATIONS FOR THE EDUCATION OF TEACHERS

If curricula are to include astronomical materials, teachers must have opportunity to acquire information in that area. However, they have not fulfilled their obligation merely by the acquisition of pertinent information. A teacher who is himself well oriented will be more likely to choose facts wisely and present them fairly than one whose own thinking is confused. Teachers who are truly concerned with the welfare of their students are obligated, therefore, to think through for themselves such matters as the relation of man to his world and to put their own thinking in order. Furthermore, teacher-educating institutions are obligated to provide opportunities for their students to achieve orientation in their own lives, to the end that they may better serve the young people whom they are to teach.

SUMMARY OF THE STUDY

Problem and plan of investigation: The broad problem of identifying changes in young people's philosophical conceptions of man's place in the universe as their information about the nature of the universe increases was attacked through a study of a specific problem, namely, to determine whether change of position on a test of opinion about man's place in the world is related to change of position on a test of information about the extent of the universe. A measure of opinion and a test of information were each prepared in two forms. A pamphlet containing astronomical information was written for use in the investigation.

Three hundred sixty-four high school physics students were tested before and after a ten-day experimental period. During this time, one group of these students studied the pamphlet without engaging in class discussion; a

Implications of the Findings 57

second group was taught as the teachers of these students deemed best for increasing comprehension of the vastness of the physical universe; the remaining students continued with their usual physics work.

The following questions deriving from the specific problem statement were asked.

Is there a change of position of the group score on the opinion scale?

Is there a relation between pre- and post-test scores of opinion?

Do the groups differ from one another with respect to their opinions at the beginning of the experiment? Do they differ from one another at its conclusion?

Is there a relation between scores on the opinion and on the information tests? Are there differences among the groups with respect to this relation?

Do the students who exhibit the greatest change of opinion also evidence the greatest change of information?

Treatment and interpretation of data: To answer these questions, certain statistics were calculated, namely,

The means of the scores

The significance of the difference between scores on pre- and post-tests of opinion

Product-moment correlations between pre- and post-tests scores of opinion, between opinion and information scores, between opinion and intelligence scores, and between changes in opinion scores and changes in information scores

The significance of the differences among groups of students with respect to these correlations

The significance of the differences between comparable pairs of correlation coefficients

The findings were these.

There was a change of position on the test of opinion in the case of the students using the pamphlet. This change was in the direction of an opinion that man is unaffected by supernatural forces.

There was a relation between pre- and post-test of opinion. Students in the control group were most likely to place themselves in the same position on the post-test as on the pretest. Students in the teacher-directed group were least likely to accept the same check-list statements of opinion on the post-test as on the pretest.

The teacher-directed group differed both at the beginning and at the end of the study from the others in that the opinions of the students in this group as evidenced by their acceptance of check-list statements were not likely to be the same as the opinions they declared themselves to hold when they indicated their positions on a self-rating scale.

There was a relation between opinion and information. The higher scores on the information test were associated with a position on the opinion test further from the opinion that man is supernaturally dominated. This relation was different for the pamphlet group than it was for the other groups. It differed on initial and on final administration of the tests for the groups in which the students were afforded an opportunity to increase their information.

There was a relation between opinion and intelligence. The higher scores on an intelligence test were associated with a position on the opinion test further from the opinion that man is supernaturally dominated. This relation was different for the pamphlet group than it was for the other groups.

The group in which the students showed the greatest change of opinion was also the group in which they showed the greatest change of information.

Inferences: These findings permit three major inferences to be drawn.

The more information young people possess about the extent of the universe, the greater is the probability that they will reject an opinion that man is governed by supernatural powers and incline toward an opposite opinion;

The more intelligent young people are those likely to reject an opinion that man is dominated by supernatural forces;

Increase in information about the extent of the universe is related to change of opinion in a direction away from a conception of supernatural domination of man.

BIBLIOGRAPHY

Burnett, R. Will. *The Opinions of Science Teachers on Some Socially Significant Issues: A Survey of Teacher Opinion and Its Implications for Teacher Education.* New York, 1940.

Campbell, Charles M. *Delusion and Belief.* Harvard University Press, 1926.

Einstein, Albert and others. *Living Philosophies.* Simon & Schuster, New York, 1931.

Fadiman, Clifton, Ed. *I Believe.* Simon & Schuster, New York, 1939.

Flanagan, John C. "General Considerations in the Selection of Test Items and a Short Method of Estimating the Product-Moment Coefficient from Data at the Tails of the Distribution." *The Journal of Educational Psychology,* Vol. 30, pp. 674-680, December, 1939.

Frank, Lawrence K. "The Task of General Education." *Frontiers of Democracy (The Social Frontier),* Vol. 3, pp. 171-173, March, 1937.

Hartshorne, Hugh, Ed. *From School to College.* Yale University Press, 1939.

Hull, Clark L. *Aptitude Testing.* World Book Co., Yonkers, N. Y., 1928.

Lorge, Irving. "Predicting Reading Difficulty of Selections for Children." *The Elementary English Review,* Vol. 16, pp. 229-233, October, 1939.

————"The Thurstone Attitude Scales. I. Reliability and Consistency of Rejection and Acceptance." *The Journal of Social Psychology,* Vol. 10, pp. 187-198, May, 1939.

Powers, S. Ralph. "The Effects of Instruction in Science on Thought, Feeling, and Action." *Teachers College Record,* Vol. 41, pp. 405-418, February, 1940.

————and Meder, Elsa M. *The Spangled Heavens.* Bureau of Educational Research in Science, Teachers College, Columbia University, New York, 1941.

Progressive Education Association, Commission on Secondary School Curriculum. *Science in General Education.* D. Appleton-Century Co., New York, 1938.

Symonds, Percival M. "Life Problems and Interests of Adolescents." *School Review,* Vol. 44, pp. 506-518, September, 1936.

Thurstone, L. L. and Chave, E. J. *The Measurement of Attitude.* University of Chicago Press, 1929.

Bei Fragen zur Produktsicherheit wenden Sie sich bitte an:
If you have any questions regarding product safety,
please contact:

Walter de Gruyter GmbH
Genthiner Straße 13
10785 Berlin
productsafety@degruyterbrill.com